SpringerBriefs in Mathematics

Series Editors

Krishnaswami Alladi
Nicola Bellomo
Michele Benzi
Tatsien Li
Matthias Neufang
Otmar Scherzer
Dierk Schleicher
Vladas Sidoravicius
Benjamin Steinberg
Yuri Tschinkel
Loring W. Tu
G. George Yin
Ping Zhang

SpringerBriefs in Mathematics showcases expositions in all areas of mathematics and applied mathematics. Manuscripts presenting new results or a single new result in a classical field, new field, or an emerging topic, applications, or bridges between new results and already published works, are encouraged. The series is intended for mathematicians and applied mathematicians.

For further volumes:
http://www.springer.com/series/10030

Markus Lohrey

The Compressed Word
Problem for Groups

Markus Lohrey

Markus Lohrey
Department for Electrical Engineering
   and Computer Science
University of Siegen
Siegen, Germany

ISSN 2191-8198   ISSN 2191-8201 (electronic)
ISBN 978-1-4939-0747-2   ISBN 978-1-4939-0748-9 (eBook)
DOI 10.1007/978-1-4939-0748-9
Springer New York Heidelberg Dordrecht London

Library of Congress Control Number: 2014934300

Mathematics Subject Classification (2010): 20F10, 20M05, 68Q17, 68Q42

Printed on acid-free paper

Springer is part of Springer Science+Business Media (www.springer.com)

Dedicated to Lynda, Andrew, and Ewan

Preface

The study of computational problems in combinatorial group theory goes back more than 100 years. In a seminal paper from 1911, Max Dehn posed three decision problems [46]: the *word problem* (called *Identitätsproblem* by Dehn), the *conjugacy problem* (called *Transformationsproblem* by Dehn), and the *isomorphism problem*. The first two problems assume a finitely generated group G (although Dehn in his paper requires a finitely presented group). For the word problem, the input consists of a finite sequence w of generators of G (also known as a finite word), and the goal is to check whether w represents the identity element of G. For the conjugacy problem, the input consists of two finite words u and v over the generators and the question is whether the group elements represented by u and v are conjugated. Finally, the isomorphism problem asks whether two given finitely presented groups are isomorphic.[1] Dehn's motivation for studying these abstract group theoretical problems came from topology. In a previous paper from 1910, Dehn studied the problem of deciding whether two knots are equivalent [45], and he realized that this problem is a special case of the isomorphism problem (whereas the question of whether a given knot can be unknotted is a special case of the word problem). In his paper from 1912 [47], Dehn gave an algorithm that solves the word problem for fundamental groups of orientable closed two-dimensional manifolds. His algorithm is nowadays known as Dehn's algorithm and can be applied to a larger class of groups (the so-called hyperbolic groups). But Dehn also realized that his three problems seem to be very hard in general. In [46], he wrote "Die drei Fundamentalprobleme für alle Gruppen mit zwei Erzeugenden ... zu lösen, scheint einstweilen noch sehr schwierig zu sein." (*Solving the three fundamental problems for all groups with two generators seems to be very difficult at the moment.*) When Dehn wrote this sentence, a formal definition of computability was still missing. Only in the first half of the 1930s, the foundations of the modern theory of computability were laid by Alonzo Church, Kurt Gödel, Emil Post, and Alan Turing; see, e.g., Chap. 1 in the textbook [40] for a brief historical outline. Another

---

[1]Implicitly, the isomorphism problem can be also found in Tietze's paper [160] from 1908.

20 years later, Pyotr Sergeyevich Novikov [136] and independently William Werner Boone [22] proved that Dehn's intuition concerning the difficulty of his problems was true: there exist finitely presented groups with an undecidable word problem (and hence an undecidable conjugacy problem). The isomorphism problem turned out to be undecidable around the same time [1, 144]. More details on the history around Dehn's three decision problems can be found in [157].

Despite these negative results, for many groups, the word problem is decidable. Dehn's result for fundamental groups of orientable closed two-dimensional manifolds was extended to one-relator groups by his student Wilhelm Magnus [121]. Other important classes of groups with a decidable word problem are automatic groups [57] (including important classes like braid groups [8], Coxeter groups, right-angled Artin groups, hyperbolic groups [68]) and finitely generated linear groups, i.e., groups of matrices over a field [145] (including polycyclic groups). This makes it interesting to study the complexity of the word problem. Computational complexity theory is a part of computer science that emerged in the mid-1960s and investigates the relation between (i) the time and memory of a computer necessary to solve a computational problem and (ii) the size of the input. For a word problem, the input size is simply the length of the finite sequence of generators. Early papers on the complexity of word problems for groups are [27–29, 145]. One of the early results in this context was that for every given $n \geq 0$ there exist groups for which the word problem is decidable but does not belong to the nth level of the Grzegorczyk hierarchy (a hierarchy of decidable problems) [27]. On the other hand, for many prominent classes of groups, the complexity of the word problem is quite low. For finitely generated linear groups Richard Jay Lipton and Yechezkel Zalcstein [108] proved that logarithmic working space (and hence polynomial time) suffices to solve the word problem. The class of automatic groups is another quite large class of groups including many classical group classes for which the word problem can be solved in quadratic time [57]. In contrast, groups with a difficult word problem seem to be quite exotic objects. The vague statement that for most groups the word problem is easy can be even made precise: Mikhail Gromov stated in [68] that a randomly chosen (according to a certain probabilistic process) finitely presented group is hyperbolic with probability 1, a proof was given by Alexander Ol'shanskiĭ in [137]. Moreover, for every hyperbolic group, the word problem can be solved in linear time. Hence, for most groups, the word problem can be solved in linear time.

One of the starting points for the work in this book was a question from the 2003 paper [94] by Ilya Kapovich, Alexei Myasnikov, Paul Eugene Schupp, and Vladimir Shpilrain: Is there a polynomial time algorithm for the word problem for the automorphism group of a finitely generated free group F? For every group G the set of all automorphisms of G together with the composition operation is a group, the automorphism group $\mathsf{Aut}(G)$ of G. Clearly, if G is finite, then also $\mathsf{Aut}(G)$ is finite. But if G is only finitely generated, then $\mathsf{Aut}(G)$ is not necessarily finitely generated; see [105] for a counterexample. In a seminal paper [135] from 1924, Jakob Nielsen proved that the automorphism group of a finitely generated free group F is finitely generated (and in fact finitely presented). This makes it interesting to study the word problem for $\mathsf{Aut}(F)$. The straightforward algorithm for this problem has an exponential running time. Saul Schleimer [151] realized in 2006 that one can easily reduce the word problem for $\mathsf{Aut}(F)$ to a variant for the word problem for F, where the input sequence is given in a succinct form by the so-called straight-line program. A straight-line program can be seen as a hierarchical description of a string. From a formal language theory point of view it is a context-free grammar $\mathbb{G}$ that generates exactly one string that we denote with $\mathsf{val}(\mathbb{G})$. The length of this string can be exponential in the size of the grammar $\mathbb{G}$. In this sense, G can be seen as a succinct description of $\mathsf{val}(\mathbb{G})$. The variant of the word problem for a group G, where the input sequence is given by a straight-line program, is the *compressed word problem for* G, which is the main topic of this book. Explicitly this problem was first studied in my paper [110] (a long version appeared as [112]). My motivation for this work came from two other papers: In 1994, Wojciech Plandowski [139] proved that one can decide in polynomial time whether two straight-line programs produce the same words; implicitly, this result can be also found in papers by Kurt Mehlhorn, Rajamani Sundar, and Christian Uhrig [127, 128] and Yoram Hirshfeld, Mark Jerrum, and Faron Moller [79, 80] that were both published the same year as Plandowski's work. One could also rephrase this result by saying that the compressed word problem for a free monoid can be solved in polynomial time. My second motivation for studying the compressed word problem is the paper [15] by Martin Beaudry, Pierre McKenzie, Pierre Péladeau, and Denis Thérien, where the authors studied the problem whether a given circuit over a finite monoid evaluates to a given monoid element. This problem can be seen as the compressed word problem for a finite monoid.[2]

Let me come back to the word problem for the automorphism group of a free group. In my paper [110] I proved that the compressed word problem for a finitely generated free group can be solved in polynomial time; the proof builds on the abovementioned result by Plandowski. Using this result, Schleimer was able to show that the word problem for $\mathsf{Aut}(F)$ (with F a finitely generated free group) can be solved in polynomial time [151], which answered the question from [94] positively.

After Schleimer's work, it turned out that the compressed word problem has applications not only for the word problem of automorphism groups but also for semidirect products and certain group extensions. This motivated the study of the compressed word problem as a topic of its own interest. The aim of this book is to give an extensive overview of known results on the compressed word problem. Some of the results in this book can be found in published articles while others appear to be new. As a computer scientist with a strong interest in group theory, I tried to make this book accessible to both mathematicians and computer scientists. In particular, I tried to explain all the necessary concepts from computer science (mainly computational complexity theory) and group theory. On the other hand,

---

[2] At this point the reader might wonder why this book is about the compressed word problems for groups only and not monoids. This is more a matter of personal taste and the fact that so far there exist more results for groups than for monoids.

to keep the book focused on the compressed word problem I had to stop at a certain point. I hope that I could find the right balance.

I want to thank my former supervisor Volker Diekert for his scientific influence. During my time in Stuttgart, he drew my interest to group theoretical decision problems and the work of Plandowski on compression. This was the starting point for my work on the compressed word problem. Many results from this book were discovered with my collaborators Niko Haubold, Christian Mathissen, and Saul Schleimer. Without their contributions, this work would not exist in its present form. I have to thank Daniel König for his thorough proof-reading. Many thanks to the referees for their numerous useful comments that helped to improve this book. Finally, I am grateful to Benjamin Steinberg for encouraging me to write this book.

Siegen, Germany   Markus Lohrey
January 2014

Contents

1 Preliminaries from Theoretical Computer Science 1
  1.1 Preliminaries from Formal Language Theory 1
  1.2 Rewriting Systems 2
  1.3 Preliminaries from Complexity Theory 4
    1.3.1 Turing Machines 4
    1.3.2 Time and Space Classes 8
    1.3.3 Reductions and Completeness 10
    1.3.4 How to Represent Input Structures 13
    1.3.5 Circuits 14
    1.3.6 Some Important Computational Problems........................ 16
    1.3.7 Randomized Computation....................................... 18
    1.3.8 The RAM Model of Computation 25

2 Preliminaries from Combinatorial Group Theory 27
  2.1 Free Groups 27
  2.2 Finitely Generated Groups and Finitely Presented Groups 28
  2.3 Tietze Transformations 30
  2.4 The Word Problem 32
  2.5 The Dehn Function and the Word Search Problem.................... 35

3 Algorithms on Compressed Words................................... 43
  3.1 Straight-Line Programs and Extensions............................. 43
  3.2 Algorithms for Words Represented by SLPs 46
  3.3 Transforming CSLPs and PCSLPs into SLPs 52
  3.4 Compressed Equality Checking 54
  3.5 2-Level PCSLPs... 63

4 The Compressed Word Problem 67
  4.1 The Compressed Word Problem and Basic Closure Properties 67
  4.2 From the Compressed Word Problem to the Word Problem 69
  4.3 The Compressed Word Problem in Finite Groups.................... 74
  4.4 The Compressed Word Problem in Free Groups 75

4.5   The Compressed Word Problem for Finitely Generated
       Linear Groups   78
4.6   The Compressed Word Problem for $\mathsf{SL}_3(\mathbb{Z})$ 79
4.7   The Compressed Word Problem for Finitely Generated
       Nilpotent Groups   80
4.8   Wreath Products: Easy Word Problem but Difficult
       Compressed Word Problem   82

5 The Compressed Word Problem in Graph Products   87
   5.1   Graph Products   87
   5.2   Trace Monoids   89
          5.2.1   General Definitions   89
          5.2.2   The Prefix and Suffix Order on Traces   91
          5.2.3   Trace Rewriting Systems   92
          5.2.4   Confluent and Terminating Trace Rewriting
                   Systems for Graph Products   93
          5.2.5   Simple Facts for Compressed Traces   96
   5.3   The Compressed Word Problem for Graph Products   98
          5.3.1   Proof of Proposition 5.35   103

6 The Compressed Word Problem in HNN-Extensions   115
   6.1   HNN-Extensions   116
   6.2   The Main Computational Problems   119
   6.3   Reducing to Britton-Reduced Sequences   121
   6.4   Reduction to a Constant Number of Stable Letters   122
   6.5   Abstracting from the Base Group K   125
   6.6   Eliminating Letters from $B_1 \cup \{t, t^{-1}\}$ 129
   6.7   Transforming G_5 into an HNN-Extension   132
   6.8   Final Step in the Proof of Theorem 6.7   133
   6.9   Amalgamated Products   134

7 Outlook   137

References   139

Acronyms and Notations   147

Index   151

Chapter 1
Preliminaries from Theoretical Computer Science

In this chapter, we briefly recall basic definitions from theoretical computer science. We start with several general notations from formal language theory (words, finite automata, word rewriting systems). The main part of the chapter is a brief introduction into basic concepts from complexity theory. The goal is to enable readers that are not familiar with complexity theory to understand the results and proofs in this book. But this chapter cannot replace a thorough introduction into the broad field of complexity theory. For this, we refer the reader to the textbooks [7, 138].

1.1 Preliminaries from Formal Language Theory

Let Γ be a set, called an *alphabet* in the following. We do not assume that Γ is finite, although most of the time it will be finite. Elements of Γ will be called *symbols*.

Definition 1.1 (finite words, length and alphabet of a word, factor, empty word).
A *finite word* over Γ is just a sequence of alphabet symbols $s = a_1 a_2 \cdots a_n$ with $n \geq 0$ and $a_1, a_2, \ldots, a_n \in \Gamma$. We can have $a_i = a_j$ for $i \neq j$. The length of this word s is $|s| = n$. For $a \in \Gamma$ let $|s|_a = |\{i \mid 1 \leq i \leq n, a_i = a\}|$ be the number of occurrences of a in s. With $\mathsf{alph}(s) = \{a \in \Gamma \mid |s|_a > 0\}$ we denote the set of symbols that occur in s, also called the *alphabet of s*. For $1 \leq i \leq n$ we write $s[i] = a_i$. For $1 \leq i \leq j \leq n$ let $s[i : j] = a_i \cdots a_j$. If $1 \leq i \leq j \leq n$ does not hold, then we set $s[i : j] = \varepsilon$. Any word of the form $s[i : j]$ is called a *factor* of s. We also use the abbreviations $s[: j] = s[1 : j]$ and $s[i :] = s[i : |s|]$. The unique word of length 0 is denoted by ε (the *empty word*).

Definition 1.2 (Γ^*, Γ^+, concatenation of words, free monoid, languages). The set of all finite words over Γ is denoted with Γ^*. Moreover, let $\Gamma^+ = \Gamma^* \setminus \{\varepsilon\}$. The *concatenation* of words $s = a_1 \cdots a_n$ and $t = b_1 \cdots b_m$ is the word $st = a_1 \cdots a_n b_1 \cdots b_m$. This is an associative operation with identity element ε.

M. Lohrey, *The Compressed Word Problem for Groups*, SpringerBriefs in Mathematics, DOI 10.1007/978-1-4939-0748-9_1, © Markus Lohrey 2014

Hence, Γ^* has the structure of a monoid, called the *free monoid* generated by Γ. A *language* is a subset of Γ^* for some finite alphabet Γ.

An important class of languages are regular languages which are defined by finite automata.

Definition 1.3 (nondeterministic finite automaton). A *nondeterministic finite automaton* (briefly *NFA*) is a tuple $\mathscr{A} = (Q, \Sigma, \delta, q_0, F)$, where

- Q is a finite set of states,
- Σ is the (finite) input alphabet,
- $\delta \subseteq Q \times \Sigma \times Q$ is the set of transitions,
- $q_0 \in Q$ is the initial state, and
- $F \subseteq Q$ is the set of final states.

We define $\hat{\delta}$ as the smallest subset of $Q \times \Sigma^* \times Q$ such that:

- $(q, \varepsilon, q) \in \hat{\delta}$ for all $q \in Q$,
- $\delta \subseteq \hat{\delta}$, and
- If $(q, u, p), (p, v, r) \in \hat{\delta}$, then also $(q, uv, r) \in \hat{\delta}$.

The *language accepted by* $\mathscr{A}$ is $L(\mathscr{A}) = \{w \in \Sigma^* \mid \exists q \in F : (q_0, w, q) \in \hat{\delta}\}$; it is the set of all words that label a path from the initial state q_0 to a final state. A *deterministic finite automaton* (briefly *DFA*) is an $\mathscr{A} = (Q, \Sigma, \delta, q_0, F)$ such that for every pair $(q, a) \in Q \times \Sigma$ there is at most one state $p \in Q$ such that $(q, a, p) \in \delta$.

By a famous theorem of Rabin and Scott, for every NFA $\mathscr{A}$ there exists a DFA $\mathscr{A}'$ such that $L(\mathscr{A}) = L(\mathscr{A}')$. There is family of NFA $\mathscr{A}_n$ such that $\mathscr{A}_n$ has n states but the smallest DFA accepting $L(\mathscr{A}_n)$ has 2^n states. When we only talk about finite automata, we refer to NFA. A language $L \subseteq \Sigma^*$ is *regular* if there is a finite automaton with $L(\mathscr{A}) = L$. The book [82] is an excellent introduction into the wide area of automata theory.

1.2 Rewriting Systems

Rewriting systems are a formalism for manipulating words or other data structures. Before we formally define word rewriting systems, we first introduce some abstract properties of binary relations; see [21, Chap. 1] for more details.

Let $\to$ be a binary relation on a set A. With $\to^*$ we denote the reflexive and transitive closure of $\to$, and $\leftrightarrow^*$ denotes the smallest equivalence relation containing $\to$. Formally, we have $a \leftrightarrow^* b$, if there exist $i \geq 1$ and elements $a_1, a_2, \ldots, a_n \in A$ such that $a_1 = a$, $a_n = b$, and $(a_i \to a_{i+1}$ or $a_{i+1} \to a_i)$ for all $1 \leq i \leq n - 1$.

Definition 1.4 (Noetherian/locally confluent/confluent relation). We say that $\rightarrow$ is *Noetherian* (or *terminating*) if there does not exist an infinite sequence $a_0 \rightarrow a_1 \rightarrow a_2 \rightarrow \cdots$ in A. The relation $\rightarrow$ is *confluent* if for all $a, b, c \in A$ with $a \rightarrow^* b$ and $a \rightarrow^* c$ there exists $d \in A$ such that $b \rightarrow^* d$ and $c \rightarrow^* d$. The relation $\rightarrow$ is *locally confluent* if for all $a, b, c \in A$ with $a \rightarrow b$ and $a \rightarrow c$ there exists $d \in A$ such that $b \rightarrow^* d$ and $c \rightarrow^* d$.

Clearly, every confluent relation is also locally confluent. Moreover, Newman's Lemma [134] states that a Noetherian and locally confluent relation is also confluent. Furthermore, if $\rightarrow$ is confluent, then for all $a, b \in A$, $a \leftrightarrow^* b$ if and only if there exists $d \in A$ with $a \rightarrow^* d$ and $b \rightarrow^* d$.

Definition 1.5 (set of irreducible words IRR($\rightarrow$), normal form NF$_\rightarrow(a)$). Let IRR($\rightarrow$) be the set of all $a \in A$ such that there does not exist $b \in A$ with $a \rightarrow b$ ("IRR" stands for "irreducible"). Assume that $\rightarrow$ is Noetherian and confluent. Then for every word $a \in A$ there exists a unique element NF$_\rightarrow(a) \in$ IRR($\rightarrow$) with $a \rightarrow^*$ NF$_\rightarrow(a)$ ("NF" stands for "normal form"). Moreover, $a \leftrightarrow^* b$ if and only if NF$_\rightarrow(a) = $ NF$_\rightarrow(b)$.

Definition 1.6 (semi-Thue system). A *semi-Thue system*[1] (or *word rewriting system*) over the alphabet Γ is just a (possibly infinite) set $R \subseteq \Gamma^* \times \Gamma^*$; see [21] for more details. We associate with R a binary relation $\rightarrow_R$ on Γ^*, also called one-step rewrite relation as follows: For all $u, v \in \Gamma^*$, $u \rightarrow_R v$ if and only if there exist $x, y \in \Gamma^*$ and $(\ell, r) \in R$ such that $u = x\ell y$ and $v = xry$. Elements (ℓ, r) are called *rules* and usually written as $\ell \rightarrow r$.

Definition 1.7 (Noetherian/confluent/locally confluent semi-Thue system, irreducible words). The semi-Thue system R is Noetherian (confluent, locally confluent) if $\rightarrow_R$ is Noetherian (confluent, locally confluent). Let IRR(R) = IRR($\rightarrow_R$) (the set of *irreducible words* with respect to R) and NF$_R(w) = $ NF$_{\rightarrow_R}(w)$ (if R is Noetherian and confluent). Thus, we have

$$\text{IRR}(R) = \Gamma^* \setminus \bigcup_{(\ell, r) \in R} \Gamma^* \ell \Gamma^*.$$

Definition 1.8 (quotient monoid Γ^*/R). Let $[u]_R = \{v \in \Gamma^* \mid u \leftrightarrow^*_R v\}$ be the equivalence class with respect to $\leftrightarrow^*_R$ containing u. Note that $\leftrightarrow^*_R$ is a congruence relation on the free monoid Γ^*: If $u_1 \leftrightarrow^*_R v_1$ and $u_2 \leftrightarrow^*_R v_2$, then $u_1 u_2 \leftrightarrow^*_R v_1 v_2$. Hence, we can define the quotient monoid Γ^*/R as follows: The set of elements of Γ^*/R is the set of equivalence classes $[u]_R$ ($u \in \Gamma^*$) and the monoid multiplication is defined by $[u]_R[v]_R = [uv]_R$.

---

[1]Named by Axel Thue, who introduced semi-Thue systems in his paper [159] from 1914.

If R is Noetherian and confluent, then every equivalence class $[u]_R$ contains a unique irreducible word $\mathsf{NF}_R(u)$. Hence, one can identify Γ^*/R with the set of irreducible words $\mathsf{IRR}(R)$, where the product of $u, v \in \mathsf{IRR}(R)$ is defined by $\mathsf{NF}_R(uv)$.

Example 1.9. The semi-Thue system $R = \{ba \to ab, ca \to ac\}$ is Noetherian and confluent. We have $\mathsf{IRR}(R) = a^*\{b, c\}^*$. The monoid $\{a, b, c\}^*/R$ is isomorphic to $\mathbb{N} \times \{a, b\}^*$ (where we take the monoid operation $+$ on $\mathbb{N}$).

For a finite Noetherian semi-Thue system R one can check whether R is locally confluent (and hence confluent by Newman's Lemma) by verifying confluence of the so-called critical pairs that result from overlapping left-hand sides of R; see [21] for more details. A survey on string rewriting systems emphasizing applications in group theory can be found in [37].

1.3 Preliminaries from Complexity Theory

We assume some familiarity with algorithms and the very basic concepts of computability; see, e.g., [7, 138] for an introduction. In this section we will only outline the basic concepts from computational complexity theory that are needed for this book.

This book is concerned with algorithmic problems in combinatorial group theory and their inherent computational difficulty. In order to make precise statements about the computational difficulty (or computational complexity) of problems, we need an abstract mathematical model of a computer. The basic computational model in computability theory and complexity theory is the Turing machine. Turing machines were introduced in Turing's seminal work [161] as an abstract model of a universal computer. There are several variants for the definition of a Turing machine. All these variants lead to the same definition of complexity classes. Here, we choose the variant of (nondeterministic and deterministic) Turing machines with a read-only input tape, a read-write working tape, and optionally a write-only output tape.

1.3.1 Turing Machines

Definition 1.10 (nondeterministic Turing machine). A *nondeterministic Turing machine* is a tuple $M = (Q, \Sigma, \Gamma, \delta, q_0, q_f, \square)$ where

- Q is a finite set of states,
- $q_0 \in Q$ is the initial state,
- $q_f \in Q$ is the final state,
- Γ is a finite tape alphabet,
- Σ is the finite input alphabet with $\Sigma \cap \{\triangleleft, \triangleright\} = \emptyset$,
- $\square \in \Gamma$ is the blank symbol, and

- $\delta \subseteq (Q \setminus \{q_f\}) \times (\Sigma \cup \{\triangleleft, \triangleright\}) \times \Gamma \times Q \times (\Gamma \setminus \{\square\}) \times \{-1, 0, 1\}^2$ is the transition relation. We call its elements instructions and require that $(q, \triangleright, a, p, b, d_1, d_2) \in \delta$ implies $d_1 \geq 0$ and $(q, \triangleleft, a, p, b, d_1, d_2) \in \delta$ implies $d_1 \leq 0$.

The nondeterministic Turing machine M is *deterministic*, if for all instructions $(q, a, b, p, c, d_1, d_2), (q, a, b, p', c', d_1', d_2') \in \delta$, we have $p = p', c = c', d_1 = d_1'$, and $d_2 = d_2'$.

The idea is that the machine M has an input tape, where the input word $w = a_1 a_2 \cdots a_n \in \Sigma^*$ is stored in cell 1 to cell n. Moreover, cell 0 (respectively, $n + 1$) of the input tape contains the left (respectively, right) end marker $\triangleright$ (respectively, $\triangleleft$). The content of the input tape is not modified by M. Moreover, the machine has a *work tape*, whose cells are indexed by the integers $\mathbb{Z}$. Each of these cells contains a symbol from Γ. At the beginning all cells contain the blank symbol $\square$, and at every step, all but finitely many cells contain the blank symbol. The machine has two heads: an input head, which is scanning a cell of the input tape, and a work head, which is scanning a cell of the work tape. Finally, the Turing machine is equipped with a finite state unit, whose current state is from Q. An instruction $(q, a, b, p, c, d_1, d_2) \in \delta$ has the following intuitive meaning: If the M is currently in state q, the input head is scanning a cell containing the symbol a, and the work tape is scanning a cell containing the symbol b, and then the machine M can do the following action: Overwrite the b in the current cell on the work tape with the non-blank symbol c (we may have $b = c$), move the input head one cell in direction d_1 (where "-1" means "one cell to the left," "0" means "stay at the same cell," and "1" means "one cell to the right"), move the work head one cell in direction d_2, and enter state q. Let us now define formally configurations and transitions of Turing machines. The following definitions all refer to the Turing machine $M = (Q, \Sigma, \Gamma, \delta, q_0, q_f, \square)$.

Definition 1.11 (configuration). A *configuration* α of the Turing machine M is a 5-tuple $\alpha = (x, q, i, w, j)$, where:

- $x \in \Sigma^*$ is the input word,
- $q \in Q$ is the current state of the Turing machine,
- $0 \leq i \leq |x| + 1$ is the cell scanned by the input head,
- $w : \mathbb{Z} \to \Gamma$ is the current content of the work tape, and there exist only finitely many $k \in \mathbb{Z}$ with $w(k) \neq \square$,
- $j \in \mathbb{Z}$ is the current cell scanned by the work head.

The *length* $|\alpha|$ of the configuration $\alpha = (x, q, i, w, j)$ is $|\alpha| = |\{k \in \mathbb{Z} \mid w(k) \neq \square\}|$. Let $\mathsf{init}(x) = (x, q_0, 0, w_\square, 0)$ be the *initial configuration* of M for input x, where $w_\square(k) = \square$ for all $k \in \mathbb{Z}$. With accept (the set of all *accepting configurations*) we denote the set of all configurations of the form (x, q_f, i, w, j).

Hence, the length $|\alpha|$ is the number of non-blank cells on the work tape. Note that the length of the input word x does not count for the length of α. This is crucial for our later definition of logspace computations. Note that $|\mathsf{init}(x)| = 0$.

For an instruction $t \in \delta$ and configurations α, β of M, we write $\alpha \vdash_t \beta$, if instruction t can be applied to configuration α and yields configuration β. Here is the formal definition.

Definition 1.12 (relation $\vdash_M$). For configurations α, β of M we write $\alpha \vdash_t \beta$ if and only if the following conditions are satisfied: $\alpha = (x, q, i_1, v, i_2)$, $\beta = (x, p, j_1, w, j_2)$ $t = (q, a_1, a_2, p, b, d_1, d_2)$, $(\triangleright x \triangleleft)[i_1 + 1] = a_1$, $v(i_2) = a_2$, $v(j) = w(j)$ for all $j \neq i_2$, $w(i_2) = b$, $j_1 = i_1 + d_1$, and $j_2 = i_2 + d_2$. We write $\alpha \vdash_M \beta$ if there exists an instruction $t \in \delta$ with $\alpha \vdash_t \beta$.

Clearly, for a nondeterministic Turing machine M, there may exist several (or no) configurations β with $\alpha \vdash_t \beta$, but the number of successor configurations of α is bounded by a constant that only depends on M. If M is deterministic, then every configuration has at most one successor configuration.

Note that if $\alpha \in$ accept, then there is no configuration β with $\alpha \vdash_M \beta$. Moreover, $\alpha \vdash_M \beta$ implies that $|\beta| - |\alpha| \in \{0, 1\}$ for all configurations α and β (since M cannot write $\square$ into a work tape cell).

Definition 1.13 (computation, $L(M)$, computably enumerable language). A *computation of M on input x* is a sequence of configurations $(\alpha_0, \alpha_1, \ldots, \alpha_m)$ with $\mathrm{init}(x) = \alpha_0$ and $\alpha_{i-1} \vdash_M \alpha_i$ for all $1 \leq i \leq m$. The computation is *successful* if $\alpha_m \in$ accept. The set

$$L(M) = \{x \in \Sigma^* \mid \exists \text{ successful computation of } M \text{ on input } x\}$$

is the language accepted (or solved) by M. A language L is called *computably enumerable* (or *recursively enumerable*) if there exists a Turing machine M with $L = L(M)$.

In the above definition of computably enumerable languages, we can restrict the Turing machine M to be deterministic. In other words, nondeterministic and deterministic Turing machines recognize the same class of languages, namely the computably enumerable languages.

Example 1.14. Consider the deterministic Turing machine

$$M = (Q, \Sigma, \Gamma, \delta, q_0, q_f, \square),$$

where $Q = \{q_0, q_1, q_2, q_3, q_f\}$, $\Sigma = \{a, b\}$, $\Gamma = \{a, b, \#\}$, and δ contains the following tuples:

1. $(q_0, \triangleright, \square, q_1, \#, 1, 1)$
2. $(q_1, x, \square, q_1, x, 1, 1)$ for $x \in \{a, b\}$
3. $(q_1, \triangleleft, \square, q_2, \#, 0, -1)$
4. $(q_2, \triangleleft, x, q_2, x, 0, -1)$ for $x \in \{a, b\}$
5. $(q_2, \triangleleft, \#, q_3, \#, -1, +1)$
6. $(q_3, x, x, q_3, \#, -1, +1)$
7. $(q_3, \triangleright, \#, q_f, \#, 0, 0)$

This machine accepts the set of all palindromes over the alphabet $\{a, b\}$, i.e., the set of words $w \in \{a, b\}^*$ such that $w[i] = w[|w| - i + 1]$ for all $1 \leq i \leq |w|$. With the first two types of instructions, we copy the input word onto the work tape. At the end of this first stage, the input head scans the right end marker $\lhd$. Then, with the instructions of type (3) and (4), we move the work tape head back to the beginning of the copy of the input word. The input head stays on the right end marker $\lhd$. The main work is done with the instructions of type (6), which compare for every $1 \leq i \leq n$ the symbols $w[i]$ and $w[|w| - i + 1]$ of the input word w.

So far, we used Turing machines only as language acceptors, similar to finite automata in automata theory. But in many contexts we want to use Turing machines to compute some possibly complex output word from the input word. For instance, we might want to output the sum or product of two input numbers. For this, we need Turing machines with output. Since such a machine should produce a single output word for a given input word, the machine should be deterministic.

Definition 1.15 (Turing machine with output). A *deterministic Turing machine with output* (also called a *transducer*) is a tuple $M = (Q, \Sigma, \Gamma, \Delta, \delta, q_0, q_f, \square)$ where

- $Q, \Sigma, \Gamma, q_0, q_f$, and $\square$ have the same meaning as for ordinary Turing machines,
- Δ is the output alphabet, and
- $\delta \subseteq (Q \setminus \{q_f\}) \times (\Sigma \cup \{\lhd, \rhd\}) \times \Gamma \times Q \times \Gamma \times \{-1, 0, 1\}^2 \times (\Delta \cup \{\varepsilon\})$ is the transition relation. We require that $(q, \rhd, a, p, b, d_1, d_2, c) \in \delta$ implies $d_1 \geq 0$ and $(q, \lhd, a, p, b, d_1, d_2, c) \in \delta$ implies $d_1 \leq 0$. Moreover, we require that for all instructions $(q, a, b, p, c, d_1, d_2, o), (q, a, b, p', c', d_1', d_2', o') \in \delta$, we have $p = p', c = c', d_1 = d_1', d_2 = d_2'$, and $o = o'$ (i.e., M is deterministic).

The intuition behind an instruction $(q, a, b, p, c, d_1, d_2, o) \in \delta$ is the same as for ordinary Turing machines, where in addition $o \in \Delta \cup \{\varepsilon\}$ is the output symbol. The idea is that the output is generated sequentially from left to right. If $o \in \Delta$, then o is appended to the prefix of the output that has been already produced. If $o = \varepsilon$, then the output is not extended in the current computation step. Formally, a configuration of M is a 6-tuple $\alpha = (x, q, i, w, j, y)$, where x, q, i, w, j have the same meaning as in Definition 1.11 and $y \in \Delta^*$ is the partial output that has been produced so far. For configurations α, β and an instruction $t \in \delta$ we write $\alpha \vdash_t \beta$ if and only if $\alpha = (x, q, i_1, v, i_2, y)$, $\beta = (x, p, j_1, w, j_2, yo)$ $t = (q, a_1, a_2, p, b, d_1, d_2, o)$, and all conditions from Definition 1.12 hold. Computations are defined analogously to ordinary Turing machines.

A Turing machine with output $M = (Q, \Sigma, \Gamma, \Delta, \delta, q_0, q_f, \square)$ computes a partial function $f_M : \Sigma^* \to \Delta^*$ in the natural way.

Definition 1.16 (function f_M computed by M). For a word $x \in \Sigma^*$ we have $f_M(x) = y$ if and only if there exists an accepting configuration $\alpha = (x, q_f, i, w, j, y)$ such that $\text{init}(x) \vdash_M^* \alpha$.

In general, the function f_M is only partially defined, since M is not guaranteed to finally reach an accepting configuration from $\text{init}(x)$.

Recall that a language L is computably enumerable if there exists a (deterministic or nondeterministic) Turing machine M with $L = L(M)$. Alternatively, one can say that L is recursively enumerable if there exists a Turing machine with output M such that L is the domain of f_M.

Definition 1.17 (computable language). A language L is *computable* (or *recursive* or *decidable*) if and only if there exists a Turing machine with output M such that for all inputs x, if $x \in L$, then $f_M(x) = 1$ and if $x \notin L$, then $f_M(x) = 0$ (in particular, f_M is total). A language which is not computable is called *undecidable*.

Intuitively, a language L is computable if there exists an algorithm such that for every input word x, the algorithm either terminates with the answer "yes, x belongs to L," or "no, x does not belong to L." It is well known that L is computable if and only if L and the complement of L are both computably enumerable.

In the following, we use the term *computational problem* or just *problem* as a synonym for language. Hence, Turing machines are devices for solving computational problems. When describing a computational problem that corresponds to a language $L \subseteq \Sigma^*$ we use a description of the following form:

input: A word x over the alphabet Σ.
question: Does x belong to L?

1.3.2   Time and Space Classes

In this section, we define several important complexity classes.

Definition 1.18 (time and space needed by a Turing machine). Let M be a Turing machine and let $c = (\alpha_0, \alpha_1, \ldots, \alpha_m)$ be a computation of M (for a certain input x). We say that c needs *time* m, and we say that c needs *space* $\max\{|\alpha_i| \mid 0 \le i \le m\}$. The Turing machine M needs time (respectively, space) at most $N \in \mathbb{N}$ on input x if every computation (not only accepting ones) of M on input x needs time (respectively, space) at most N.

Definition 1.19 (f-time bounded, f-space bounded). Let $f : \mathbb{N} \to \mathbb{N}$ be a monotonically increasing function. We say that M is f-*time bounded* if for every input x, M needs time at most $f(|x|)$. We say that M is f-*space bounded* if for every input x, M needs space at most $f(|x|)$.

For example, the Turing machine from Example 1.14 is $(3n + 4)$-time bounded.

Of course, the above definitions can be also used for Turing machines with output.

Definition 1.20 (DTIME, NTIME, DSPACE, NSPACE). We define the following complexity classes:

$$\mathsf{DTIME}(f) = \{L(M) \mid M \text{ is deterministic and } f\text{-time bounded}\}$$

NTIME$(f) = \{L(M) \mid M$ is nondeterministic and f-time bounded$\}$

DSPACE$(f) = \{L(M) \mid M$ is deterministic and f-space bounded$\}$

NSPACE$(f) = \{L(M) \mid M$ is nondeterministic and f-space bounded.$\}$

For a class $\mathscr{C}$ of monotonically increasing functions on $\mathbb{N}$ we write DTIME$(\mathscr{C})$ for $\bigcup_{f \in \mathscr{C}}$ DTIME(f) and similarly for NTIME, DSPACE, and NSPACE.

Trivially, DTIME$(f) \subseteq$ NTIME(f) and DSPACE$(f) \subseteq$ NSPACE(f). Some other inclusions are:

- DSPACE$(\mathscr{O}(f)) =$ DSPACE(f) and NSPACE$(\mathscr{O}(f)) =$ NSPACE(f)
- NTIME$(f) \subseteq$ DSPACE(f)
- NSPACE$(f) \subseteq$ DTIME$(2^{\mathscr{O}(f)})$ if $f \in \Omega(\log n)$.

Moreover, a famous theorem of Savitch states that NSPACE$(f) \subseteq$ DSPACE(f^2) if $f \in \Omega(\log n)$. Hence, nondeterministic Turing machines can be simulated by deterministic Turing machines with a quadratic blow up in space.

Definition 1.21 (L, P, NP, PSPACE). Some important abbreviations are the following, where as usual $\mathbb{N}[x]$ denotes the class of all polynomials with coefficients from $\mathbb{N}$:

$$\mathbf{L} = \mathsf{DSPACE}(\log(n)) \tag{1.1}$$

$$\mathbf{NL} = \mathsf{NSPACE}(\log(n)) \tag{1.2}$$

$$\mathbf{P} = \bigcup_{f \in \mathbb{N}[n]} \mathsf{DTIME}(f) \tag{1.3}$$

$$\mathbf{NP} = \bigcup_{f \in \mathbb{N}[n]} \mathsf{NTIME}(f) \tag{1.4}$$

$$\mathbf{PSPACE} = \bigcup_{f \in \mathbb{N}[n]} \mathsf{DSPACE}(f) \overset{*}{=} \bigcup_{f \in \mathbb{N}[n]} \mathsf{NSPACE}(f). \tag{1.5}$$

The identity (*) in (1.5) is a consequence of Savitch's theorem. In (1.1) and (1.2) we do not have to specify the base of the logarithm, since changing the base only involves a multiplicative factor. So, **L** (respectively, **NL**) is the class of all problems that can be decided by a deterministic (respectively, nondeterministic) Turing machine in logarithmic space, **P** (respectively, **NP**) is the class of all problems that can be decided by a deterministic (respectively, nondeterministic) Turing machine in polynomial time, and **PSPACE** is the class of all problems that can be decided by a (deterministic or nondeterministic) Turing machine in polynomial space. The class **P** is often identified with the class of those problems that can be solved efficiently. The following inclusions hold:

$$\mathbf{L} \subseteq \mathbf{NL} \subseteq \mathbf{P} \subseteq \mathbf{NP} \subseteq \mathbf{PSPACE}. \tag{1.6}$$

But none of these inclusions is known to be strict (and the strictness of each inclusion is a major open problem in complexity theory), although the so- called space hierarchy theorem implies $\mathbf{NL} \subsetneq \mathbf{PSPACE}$; hence one of the inclusions must be proper.

Definition 1.22 (coC). For a complexity class **C** we denote with **coC** the set of all complements of languages from **C**.

Hence, $\mathbf{C} = \mathbf{coC}$ means that **C** is closed under complement. It is easy to show that deterministic time and space classes (like **L**, **P**, **PSPACE**) are closed under complement. A famous result shown independently by Immerman and Szelepcsényi in 1985 states that $\mathbf{NL} = \mathbf{coNL}$. Whether $\mathbf{NP} = \mathbf{coNP}$ is a major open problem in complexity theory.

1.3.3 Reductions and Completeness

In this section we will introduce the important notions of reductions and completeness for a complexity class. Via completeness, one can identify the most difficult problems within a complexity class. First we have to introduce some resource bounded classes of transducers.

Definition 1.23 (polynomial time transducer, logspace transducer). A *polynomial time transducer* is a Turing machine M with output that computes a total function f_M and that is $p(n)$-time bounded for a polynomial $p(n)$. A *logspace transducer* is a Turing machine M with output that computes a total function f_M and that is $\log(n)$-space bounded.

Example 1.24. The binary representation of the sum of two natural numbers that are given in binary representation can be computed with a logspace transducer that follows the standard school method for addition. If u and v are the binary representations of the two input numbers (let us assume that the least significant bit is left), then the machine has to store a position $1 \le p \le \max\{|u|, |v|\}$. This position can be stored in binary representation and hence needs only logarithmically many bits. Moreover, a carry bit is stored. In each step, the bits at position p in u and v and the carry bit are added, a new output bit is produced, and a new carry bit is stored. Moreover, the position p is incremented. Incrementing a number in binary representation is possible in logarithmic space.

Also the product of two binary encoded natural numbers can be computed by a logspace transducer; this requires a little bit more work than for sum. Finally, also the integer part of the quotient of two binary encoded natural numbers can be computed in logspace [35, 76]. This solved a long-standing open problem.

Example 1.25. Another class of functions that can be computed with logspace transducers are free monoid morphisms. Let $f : \Sigma^* \to \Gamma^*$ be a morphism between free monoids. A logspace transducer that computes f simply reads the input word x from left to right, and for every symbol a in x, it appends $f(a)$ to the output.

Using polynomial time transducers and logspace transducers we can refine the classical many-one reducibility from computability theory.

Definition 1.26 ($\leq_m^P$, $\leq_m^{\log}$). Let $K \subseteq \Sigma^*$ and $L \subseteq \Delta^*$ be languages. We write $K \leq_m^P L$ (K is polynomial time many-one reducible to L) if there exists a polynomial time transducer M such that for all $x \in \Sigma^*$ we have: $x \in K$ if and only if $f_M(x) \in L$. We write $K \leq_m^{\log} L$ (K is logspace many-one reducible to L) if there exists a logspace transducer M such that for all $x \in \Sigma^*$ we have $x \in K$ if and only if $f_M(x) \in L$.

Since a logspace transducer works in polynomial time (this follows from the same argument that shows $\mathbf{L} \subseteq \mathbf{P}$), $K \leq_m^{\log} L$ implies $K \leq_m^P L$. Moreover, $L_1 \leq_m^P L_2 \leq_m^P L_3$ implies $L_1 \leq_m^P L_3$, i.e., $\leq_m^P$ is transitive: Simply compose two polynomial time transducers. This simple approach does not work for logspace transducers: If M_1 and M_2 are logspace transducers, then for an input x, the length of $f_{M_1}(x)$ can be polynomial in the length of x. Hence, we cannot store the word $f_{M_1}(x)$ as an intermediate result. Nevertheless, $\leq_m^{\log}$ is a transitive relation. The proof is neither obvious nor very difficult. Transitivity of $\leq_m^{\log}$ is very useful. In order to construct a logspace transducer for a certain task, it allows to split this task into several (constantly many) subtasks and to design a logspace transducer for each subtask.

The class $\mathbf{L}$ is closed under $\leq_m^{\log}$: If $L \in \mathbf{L}$ and $K \leq_m^{\log} L$, then also $K \in \mathbf{L}$. The same holds for all complexity classes in (1.6). Similarly, $\mathbf{P}$ and all complexity class above $\mathbf{P}$ in (1.6) are closed under $\leq_m^P$.

The following two types of reducibilities are weaker than $\leq_m^{\log}$ and $\leq_m^P$, respectively.

Definition 1.27 ($\leq_{bc}^P$, $\leq_{bc}^{\log}$). Let $K \subseteq \Sigma^*$ and $L \subseteq \Delta^*$ be languages. We write $K \leq_{bc}^P L$ (K is polynomial time bounded conjunctively reducible to L) if there exist a constant $c \in \mathbb{N}$ and a polynomial time transducer M such that $f_M(\Sigma^*) \subseteq (\# \Delta^*)^c$ (where $\# \notin \Delta$) and for all $x \in \Sigma^*$ we have $x \in K$ if and only if $\bigwedge_{i=1}^c y_i \in L$, where $f_M(x) = \# y_1 \# y_2 \cdots \# y_c$. We write $K \leq_{bc}^{\log} L$ (K is logspace bounded conjunctively reducible to L) if there exist a constant $c \in \mathbb{N}$ and a logspace transducer M such that $f_M(\Sigma^*) \subseteq (\# \Delta^*)^c$ (where $\# \notin \Delta$) and for all $x \in \Sigma^*$ we have $x \in K$ if and only if $\bigwedge_{i=1}^c y_i \in L$, where $f_M(x) = \# y_1 \# y_2 \cdots \# y_c$.

The relation $\leq_{bc}^P$ (respectively, $\leq_{bc}^{\log}$) has the same closure properties that we mentioned for $\leq_m^P$ (respectively, $\leq_m^{\log}$) above.

In this book, we will also need a fifth type of reducibility that is weaker than $\leq_{bc}^P$. It is called *polynomial time Turing-reducibility* and is defined via *oracle Turing machines*.

Definition 1.28 (oracle Turing machine). Let us fix a language $L \in \Delta^*$. An *oracle Turing machine with oracle access* to L is a tuple $M = (Q, \Sigma, \Gamma, \Delta, \delta, q_0, q_f, \square)$ that satisfies all the conditions of a Turing machine with output (Definition 1.15). In addition, Q contains three distinguished states $q_?, q_Y$, and q_N such that δ contains no outgoing transitions from state $q_?$. Instead, we enforce that for a configuration $\alpha = (q_?, i, w, j, y)$ (hence, $y \in \Delta^*$ is the output produced so far), the next configuration is $(q_Y, i, w, j, \varepsilon)$ if $y \in L$ and $(q_N, i, w, j, \varepsilon)$ if $y \notin L$.

Intuitively this means that M can compute words over Δ (the so-called oracle queries). When it enters the state $q_?$ (the query state), the computed oracle query is send to an L-oracle. This oracle answers instantaneously, whether the oracle query belongs to L or not. Hence, membership in the set L can be tested for free. One could also say that we enhance the machine M with a black-box procedure that tests membership in the set L.

In Definition 1.28, we do not care whether L is computable or not. Oracle Turing machines can be deterministic or nondeterministic, but we will only use deterministic ones. All definitions that we gave for ordinary Turing machines (e.g., the accepted language $L(M)$, $f(n)$-time bounded, $f(n)$-space bounded) are defined analogously to ordinary Turing machines. Now we can define polynomial time Turing-reducibility.

Definition 1.29 ($\leq_T^P$). For languages L and K we write $K \leq_T^P L$ (K is polynomial time Turing-reducible to K) if there exists a deterministic polynomially time bounded oracle Turing machine M with oracle access to L such that $K = L(M)$.

In other words, K can be decided in polynomial time using an oracle for the language L. Clearly, $K \leq_m^P L$ implies $K \leq_T^P L$. The converse is false: For every language $L \subseteq \{0, 1\}^*$, we have $L \leq_T^P (\{0, 1\}^* \setminus L)$, but there are languages such that $L \leq_m^P (\{0, 1\}^* \setminus L)$ does not hold (e.g., the halting problem).

Let $\leq$ be one of our reducibility notions, and let $K, L_1, \ldots, L_n$ be languages. When writing $K \leq \{L_1, \ldots, L_n\}$ we mean that $K \leq \#L_1 \# L_2 \# \cdots L_n$.

Using our reducibility relations we can finally define complete problems for complexity classes. Complete problems for a complexity class $\mathscr{C}$ are the most difficult problems for the class $\mathscr{C}$. In some sense they completely capture the class $\mathscr{C}$.

Definition 1.30 ($\mathscr{C}$-hard language, $\mathscr{C}$-complete language). Let $\mathscr{C}$ be a complexity class and let $\leq$ be one of the above reducibility relations ($\leq_m^{\log}, \leq_P^m, \leq_{bc}^{\log}, \leq_{bc}^P$, or $\leq_P^T$). A language L is $\mathscr{C}$-hard with respect to $\leq$ if $K \leq L$ for every $K \in \mathscr{C}$ (L is at least as difficult as any problem in $\mathscr{C}$). The language L is $\mathscr{C}$-complete with respect to $\leq$ if $L \in \mathscr{C}$ and L is $\mathscr{C}$-hard with respect to $\leq$. Whenever we say that L is $\mathscr{C}$-complete, we implicitly refer to logspace many-one reducibility $\leq_m^{\log}$.

Complete problems for $\mathscr{C}$ are in a sense the most difficult problems from $\mathscr{C}$. For this, it is important that the reducibility used for the definition of completeness is not too weak. For instance, any language $L \subseteq \{0, 1\}^*$ with $\emptyset \neq L \neq \{0, 1\}^*$ is **P**-complete with respect to $\leq_P^m$.

For many complexity classes, very natural complete problems are known, Sect. 1.3.6 below mentions some of them. The book [58] contains a list of several hundred **NP**-complete problems, and [67] contains a similar list of **P**-complete problems. **NP**-complete problems are often called intractable, because of the following observation: If L is **NP**-complete, then $L \in \mathbf{P}$ if and only if $\mathbf{P} = \mathbf{NP}$. Analogous equivalences hold for other complexity classes too: For instance, if L is **P**-complete, then $L \in \mathbf{NL}$ if and only if $\mathbf{NL} = \mathbf{P}$.

1.3.4 How to Represent Input Structures

We should say a few words about the input representation. The input for a Turing machine is a word over some alphabet Σ. But quite often, we want to solve algorithmic problems, where the input is not a word, but, for instance, a finite graph, or a finite relational structure, or even another Turing machine. If we want to solve such a problem with a Turing machine, we have to encode the input data types by words. This is certainly possible. For instance, a directed graph on n nodes can be represented by its adjacency matrix. This matrix can be encoded by a word over the alphabet $\{0, 1, \#\}$ by listing all matrix rows separated by $\#$. Of course, there are other reasonable representations of graphs. For instance, a directed graph on n vertices $1, \ldots, n$ can be represented by concatenating for $1 \leq i \leq n$ all words $0^i \# 0^{j_1} \# 0^{j_2} \cdots \# 0^{j_k}$ where $j_1, \ldots, j_k$ is a list of all successor nodes of node i. This encoding is based on adjacency lists. Strictly speaking the precise word encoding of the input data has to be fixed for a computational problem. Quite often this is not done explicitly, but some canonical encoding is assumed. If two different input encodings can be transformed into each other efficiently, say by a logspace transducer, then the complexity of the problem does not depend on which of the two encodings is chosen (at least if we only consider classes containing **L**). For instance, the word encodings of directed graphs based on adjacency matrices and adjacency lists can be converted into each other by a logspace transducer.

The two important representations of numbers, which cannot be transformed into each other by logspace transducers, are the unary and binary representation of numbers. The unary representation of the number $n \in \mathbb{N}$ is just the word 0^n. A polynomial time transducer cannot convert the binary encoding of the number n into its unary encoding, simply because the latter is exponentially longer than the former, but in polynomial time only polynomially many output symbols can be produced. Hence, for computational problems, where the input consists of numbers, it is important to specify the encoding of numbers (unary or binary encoding). One also says that the binary encoding of numbers is more succinct than the unary encoding. The topic of succinctness is central for this book. In the next section, we will see a general framework for succinct representations based on circuits.

A final remark in this context is that for most natural encodings, it is very easy to check for a Turing machine whether a given input word is a valid encoding. For instance, whether a word over the alphabet $\{0, 1, \#\}$ is the word encoding of a graph as described above can be verified by a deterministic Turing machine in logarithmic space.

1.3.5 Circuits

Circuits are a central topic in this book. They allow to represent large objects in a succinct way. In this section, we introduce a general definition of circuits. Later, we will see several concrete instantiations of circuits.

Let us fix a countable (possibly finite) set $\mathscr{F}$ of function symbols, where every $f \in \mathscr{F}$ has a certain arity $n_f \in \mathbb{N}$. In case $n_f = 0$, f is a constant symbol. We assume that $\mathscr{F}$ contains at least one constant symbol.

Definition 1.31 ($\mathscr{F}$-expressions with variables). Let V be a finite set of *variables*. The set $\mathscr{E}(\mathscr{F}, V)$ of $\mathscr{F}$-*expressions* with variables from V is the smallest set with respect to inclusion such that:

- $V \subseteq \mathscr{E}(\mathscr{F}, V)$
- If $f \in \mathscr{F}$ and $e_1, \ldots, e_{n_f} \in \mathscr{E}(\mathscr{F}, V)$, then also $f(e_1, \ldots, e_{n_f}) \in \mathscr{E}(\mathscr{F}, V)$.

The set $\mathsf{Var}(e)$ of variables that occur in the expression e is defined as follows:

- $\mathsf{Var}(v) = \{v\}$ for $v \in V$.
- $\mathsf{Var}(f(e_1, \ldots, e_{n_f})) = \bigcup_{1 \leq i \leq n_f} \mathsf{Var}(e_i)$.

Circuits are a succinct representation of expressions without variables. The idea is to represent a subexpression that appears several times only once.

Definition 1.32 (circuit). An $\mathscr{F}$-*circuit* is a triple $\mathscr{C} = (V, S, \mathrm{rhs})$ such that:

- V is a finite set of variables (or circuit gates).
- $S \in V$ is the start variable.
- $\mathrm{rhs} : V \to \mathscr{E}(\mathscr{F}, V)$ (rhs stands for "right-hand side") assigns to each variable an expression such that the relation

$$\mathsf{graph}(\mathscr{C}) = \{(A, B) \in V \times V \mid A \in \mathsf{Var}(\mathrm{rhs}(B))\} \qquad (1.7)$$

is acyclic. The reflexive and transitive closure of the relation $\mathsf{graph}(\mathscr{C})$ (which is a partial order on V) is the *hierarchical order* of $\mathscr{C}$. The length (number of edges) of a longest path in the graph $(V, \mathsf{graph}(\mathscr{C}))$ is the *height* of $\mathscr{C}$.

In order to evaluate a circuit we need an interpretation for the function symbols from $\mathscr{F}$. This leads to the notion of an $\mathscr{F}$-algebra.

Definition 1.33 ($\mathscr{F}$-algebra). An $\mathscr{F}$-algebra is a pair $\mathscr{A} = (A, I)$, where A is a set and I maps every $f \in \mathscr{F}$ to a function $I(f) : A^{n_f} \to A$. We usually suppress the mapping I and simply write $(A, (f)_{f \in \mathscr{F}})$ for (A, I). Here, f is identified with the function $I(f)$.

Definition 1.34 (evaluation of an $\mathscr{F}$-circuit in an $\mathscr{F}$-algebra). Fix an $\mathscr{F}$-algebra $\mathscr{A} = (A, I)$ and an $\mathscr{F}$-circuit $\mathscr{C} = (V, S, \mathrm{rhs})$. We define the evaluation mapping $\mathrm{val}_{\mathscr{C}}^{\mathscr{A}} : \mathscr{E}(\mathscr{F}, V) \to A$ inductively as follows:

- If $e = X \in V$, then $\mathrm{val}_{\mathscr{C}}^{\mathscr{A}}(e) = \mathrm{val}_{\mathscr{C}}^{\mathscr{A}}(\mathrm{rhs}(X))$.
- If $e = f(e_1, \ldots, e_{n_f})$, then $\mathrm{val}_{\mathscr{C}}^{\mathscr{A}}(e) = f(\mathrm{val}_{\mathscr{C}}^{\mathscr{A}}(e_1), \ldots, \mathrm{val}_{\mathscr{C}}^{\mathscr{A}}(e_{n_f})) \in A$.

The fact that $(V, \mathsf{graph}(\mathscr{C}))$ is acyclic ensures that the mapping $\mathrm{val}_{\mathscr{C}}^{\mathscr{A}}$ is well defined. Finally, we define $\mathrm{val}^{\mathscr{A}}(\mathscr{C}) = \mathrm{val}_{\mathscr{C}}^{\mathscr{A}}(S)$.

The $\mathscr{F}$-algebra $\mathscr{A}$ will be always clear from the context. Hence, we will omit the superscript $\mathscr{A}$ in the mappings $\mathrm{val}_{\mathscr{C}}^{\mathscr{A}}$ and $\mathrm{val}^{\mathscr{A}}$. We will also talk about $\mathscr{A}$-expressions and $\mathscr{A}$-circuits (or circuits over $\mathscr{A}$) instead of $\mathscr{F}$-expressions and $\mathscr{F}$-circuits. Finally, if also the circuit $\mathscr{C}$ is clear from the context, we will also omit the index $\mathscr{C}$ in the mapping $\mathrm{val}_{\mathscr{C}}(e)$.

If a Turing machine wants to receive a circuit $\mathscr{C} = (V, S, \mathrm{rhs})$ as input, the circuit has to be encoded by a word. One can identify the variables with numbers from $\{1, \ldots, n\}$ and each of these numbers can be binary encoded. Moreover, one can assume $S = 1$. Then, to encode the whole circuit, we first specify the number n (number of variables) and then list all right-hand sides $\mathrm{rhs}(1), \ldots, \mathrm{rhs}(n)$. In order to encode an expression $e \in \mathscr{E}(\mathscr{A}, V)$ we have to specify how to encode the operation symbols $f \in \mathscr{F}$. If $\mathscr{F}$ is finite, we can simply include the set $\mathscr{F}$ into the set of input symbols for the Turing machine. If $\mathscr{F}$ is countably infinite, we can assume that $\mathscr{F} = \mathbb{N}$ and take the binary encoding of f as the code for the function symbol f. In order to be flexible, let us fix a function $\mathsf{size} : \mathscr{F} \to \mathbb{N}$. In case $\mathscr{F}$ is finite, we assume that $\mathsf{size}(f) = 1$ for every $f \in \mathscr{F}$. This is reasonable, since a symbol from $\mathscr{F}$ can be specified in constant space.

Definition 1.35 (size of a circuit). Having fixed a size function, we define the size $|e|$ of an expression $e \in \mathscr{E}(\mathscr{F}, V)$ inductively as follows:

- If $e \in V$, then $|e| = 1$.
- If $e = f(e_1, \ldots, e_{n_f})$, then $|e| = \mathsf{size}(f) + \sum_{i=1}^{n_f} |e_i|$.

Finally, we define the *size* $|\mathscr{C}|$ of the $\mathscr{F}$-circuit $\mathscr{C} = (V, S, \mathrm{rhs})$ as

$$|\mathscr{C}| = \sum_{X \in V} |\mathrm{rhs}(X)|.$$

Definition 1.36 (circuit in normal form). A circuit $\mathscr{C} = (V, S, \mathrm{rhs})$ is in normal form if for every $X \in V$, we have $\mathrm{rhs}(X) = f(X_1, \ldots, X_{n_f})$ for some $f \in \mathscr{F}$ and $X_1, \ldots, X_{n_f} \in V$.

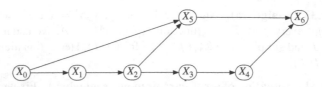

Fig. 1.1 The graph of the circuit from Example 1.37

By introducing additional variables we can transform in polynomial time a given circuit $\mathscr{C}$ into a circuit $\mathscr{D}$ in normal form such that $\mathsf{val}(\mathscr{C}) = \mathsf{val}(\mathscr{D})$.

Example 1.37. Let $\mathscr{F} = \{+, \times, 1\}$, where $n_+ = n_\times = 2$ and $n_1 = 0$. Moreover, consider the $\mathscr{F}$-algebra $\mathscr{A} = (\mathbb{Z}, I)$, where $I(+)$ is the addition operation, $I(\times)$ is the multiplication operation, and $I(1) = 1$. Finally, define the $\mathscr{F}$-circuit

$$\mathscr{C} = (\{X_0, X_1, X_2, X_3, X_4, X_5, X_6\}, X_6, \mathrm{rhs}),$$

where the right-hand side mapping rhs is defined as follows (we use infix notation for $+$ and $\times$):

$$\mathrm{rhs}(X_0) = 1$$
$$\mathrm{rhs}(X_1) = X_0 + X_0$$
$$\mathrm{rhs}(X_i) = X_{i-1} \times X_{i-1} \text{ for } 2 \leq i \leq 4$$
$$\mathrm{rhs}(X_5) = X_2 + X_0$$
$$\mathrm{rhs}(X_6) = X_5 + X_4.$$

We obtain

$$\mathsf{val}_{\mathscr{C}}(X_0) = 1, \quad \mathsf{val}_{\mathscr{C}}(X_1) = 2, \quad \mathsf{val}_{\mathscr{C}}(X_2) = 4, \quad \mathsf{val}_{\mathscr{C}}(X_3) = 16,$$
$$\mathsf{val}_{\mathscr{C}}(X_4) = 256, \quad \mathsf{val}_{\mathscr{C}}(X_5) = 5, \quad \mathsf{val}_{\mathscr{C}}(X_6) = 261$$

and hence $\mathsf{val}(\mathscr{C}) = 261$. The circuit $\mathscr{C}$ is in normal form and has size 19. The graph of $\mathscr{C}$ is shown in Fig. 1.1.

1.3.6   Some Important Computational Problems

In this section, we present some important algorithmic problems. In particular, we present complete problems for the complexity classes **L**, **NL**, **P**, **NP**, and **PSPACE**.

Example 1.38 (graph accessibility). The following problem is known as the *graph accessibility problem*:

input: A directed graph G and two nodes s and t.
question: Is there a directed path from s to t in G?

The graph accessibility problem belongs to **NL** and is actually **NL**-complete. The *undirected graph accessibility problem* is the following restriction of the graph accessibility problem:

input: An undirected graph G and two nodes s and t.
question: Is there a path from s to t in G?

The undirected graph accessibility problem belongs to **L**. This is a very difficult result which was proved in 2004 by Reingold [146]. Actually, the undirected graph accessibility problem is **L**-complete. In order to make this statement nontrivial, one has to work with a stronger reducibility than logspace reducibility (the so-called **DLOGTIME**-reducibility).

Example 1.39 (PRIMES). The following problem PRIMES belongs to **P**.

input: A word $w \in \{0, 1\}^*$.
question: Is w the binary representation of a prime number?

This was shown in 2004 by Agrawal et al. [4]. Note that is easy to test whether for an input word $w = 0^n$ (unary representation of n), n is a prime number. This latter problem actually belongs to **L**. It is not known whether PRIMES is **P**-complete.

Example 1.40 (circuit value problem). A *Boolean circuit* is a $\mathscr{B}$-circuit (in the sense of Definition 1.32), where $\mathscr{B} = (\{0, 1\}, \wedge, \vee, \neg, 0, 1)$ is the Boolean algebra ($\wedge$ denotes the binary Boolean AND-function, $\vee$ denotes the binary Boolean OR-function, and $\neg$ denotes the unary Boolean NOT-function). The *circuit value problem* is the following problem:

input: A Boolean circuit C.
question: Does the output gate of C evaluate to 1?

Ladner proved that the circuit value problem is **P**-complete [101]. Moreover, the *monotone circuit value problem*, which is the restriction of the circuit value problem to monotone Boolean circuits (Boolean circuits that do not contain the $\neg$-operator on right-hand sides), is **P**-complete [63] as well.

Example 1.41 (SAT). A Boolean formula is an expression from $\mathscr{E}(\mathscr{B}, V)$ (in the sense of Definition 1.31), where $\mathscr{B}$ is the Boolean algebra from the circuit value problem and V is a finite set of variables. Such a Boolean formula F is satisfiable if the variables from V can be replaced by truth values (0 or 1) in such a way that F evaluates to 1. SAT is the following problem:

input: A Boolean formula F.
question: Is F satisfiable?

SAT was shown to be **NP**-complete by Cook [39] and independently by Levin [104]. SAT was actually the first natural problem that was shown to be **NP**-complete.

Example 1.42 (SUBSETSUM). SUBSETSUM is the following problem:

input: A finite set $W \subseteq \mathbb{N}$ and $t \in \mathbb{N}$, where all numbers are encoded in binary.
question: Is there a subset $V \subseteq W$ such that $\sum_{w \in V} w = t$?

SUBSETSUM is **NP**-complete as well.

Example 1.43 (universality problem for NFAs). The universality problem for non-deterministic finite automata is the following problem:

input: A NFA A over the alphabet $\{0, 1\}$.
question: Does $L(A) = \{0, 1\}^*$ hold, i.e., does A accept all input words?

This problem was shown to be **PSPACE**-complete in [5].

1.3.7 Randomized Computation

Let M be a nondeterministic Turing machine, and let x be an input word for M. The machine M generates a *computation tree* for input x. This is a rooted tree, where every node is labeled with a configuration. The root is labeled with the initial configuration $\mathsf{init}(x)$, and if a node v is labeled with a configuration α, then for every configuration β with $\alpha \vdash_M \beta$, v has a β-labeled child. Assume now that this computation tree is finite, which means that M does not have an infinite computation on input x. We may view the computation tree as a stochastic process, where at each node v with children $v_1, \ldots, v_k$ one of the nodes $v_1, \ldots, v_k$ is chosen with probability $1/k$. We can now compute the probability $\mathsf{Prob}[M$ accepts $x]$ that M accepts the input x. This new viewpoint allows us to define several randomized complexity classes.

Definition 1.44 (RP). A language L belongs to the class **RP** (*randomized polynomial time*) if there exists a nondeterministic polynomial time bounded Turing machine M such that for every input x we have:

- If $x \notin L$, then $\mathsf{Prob}[M$ accepts $x] = 0$.
- If $x \in L$, then $\mathsf{Prob}[M$ accepts $x] \geq 1/2$.

The choice of the probability $1/2$ in the second case is not crucial. If the acceptance probability in case $x \in L$ is $1/p(n)$ (where $n = |x|$ is the input length), i.e., the error probability is $1 - 1/p(n)$, then by running the algorithm $n \cdot p(n)$ times, we can reduce the error probability to $(1 - 1/p(n))^{n \cdot p(n)}$, which is smaller than $(1/2)^n$ for n large enough. It is important here that the different runs of M on input x are mutually independent.

Note that a language L belongs to the class **coRP** if there exists a nondeterministic polynomial time bounded Turing machine M such that for every input x we have:

- If $x \in L$, then $\mathsf{Prob}[M$ accepts $x] = 1$.
- If $x \notin L$, then $\mathsf{Prob}[M$ accepts $x] \leq 1/2$.

The class **RP** ∩ **coRP** is also called **ZPP** (*zero-error probabilistic polynomial time*). Note that **P** ⊆ **ZPP**.

Definition 1.45 (BPP). A language L belongs to the class **BPP** (*bounded error probabilistic polynomial time*) if there exists a nondeterministic polynomial time bounded Turing machine M and a constant ϵ such that for every input x we have:

- if $x \notin L$, then Prob[M accepts x] $\leq 1/2 - \epsilon$.
- if $x \in L$, then Prob[M accepts x] $\geq 1/2 + \epsilon$.

The constant ϵ (the probability gap) can be made larger by probability amplification. For this the machine runs several times (say $2k - 1$ times) on input x and outputs by majority, i.e., the machine accepts if the machine accepts at least k times. Chernoff's bound can be used for the analysis of this new machine.

Note that **RP** ∪ **coRP** ⊆ **BPP**. The classes **RP**, **coRP**, and **BPP** are closed under polynomial time many-one reductions: If $L \in$ **RP** and $K \leq^P_m L$, then also $K \in$ **RP** and similarly for **coRP** and **BPP**. Since it is not known whether **RP** is closed under complement, it is not clear whether **RP** is closed under polynomial time Turing-reductions. But we will need the following.

Lemma 1.46. *Let* $L \in$ **RP** *(respectively,* **coRP***) and* $K \leq^P_{bc} L$. *Then also* $K \in$ **RP** *(respectively,* $K \in$ **coRP***).*

Proof. Let $K \subseteq \Sigma^*$ and $L \subseteq \Delta^*$. By the definition of $\leq^P_{bc}$, there exist a constant c and a polynomial time transducer T such that $f_T(\Sigma^*) \subseteq (\#\Delta^*)^c$ and for all $x \in \Sigma^*$ we have $x \in K$ if and only if $\bigwedge_{i=1}^c y_i \in L$, where $f_T(x) = \#y_1\#y_2 \cdots \#y_c$.

First, assume that $L \in$ **RP**. Hence, there exists a nondeterministic polynomial time Turing machine M such that for all $y \in \Delta^*$, (i) if $y \notin L$, then Prob[M accepts y] $= 0$ and (ii) if $y \in L$, then Prob[M accepts y] $\geq 1/2$. Consider the nondeterministic polynomial time Turing machine M' that behaves for all $x \in \Sigma^*$ as follows:

- From x, compute in polynomial time the word $f_T(x) = \#y_1\#y_2 \cdots \#y_c$.
- Run the machine M on each of the input words y_i. These computations must be mutually independent.
- If each of these c many computations is accepting, then M' accepts x; otherwise, M' rejects x.

Let us compute the probability that M' accepts x. Since the computations are mutually independent, we have

$$\text{Prob}[M' \text{ accepts } x] = \prod_{i=1}^c \text{Prob}[M \text{ accepts } y_i]. \tag{1.8}$$

If $x \notin K$, then there exists at least one y_i $(1 \leq i \leq c)$ such that $y_i \notin L$. Hence, Prob[M accepts y_i] $= 0$, which implies Prob[M' accepts x] $= 0$ with (1.8). If $x \in K$, then $y_i \in L$ for all $1 \leq i \leq c$. Hence, Prob[M accepts y_i] $\geq 1/2$ for all $1 \leq i \leq c$. Hence, Prob[M' accepts x] $\geq (1/2)^c$, which is a constant (since c is a constant). Using probability amplification, we can increase this constant to at least $1/2$.

Now, assume that $L \in$ **coRP**. Hence, there exists a nondeterministic polynomial time Turing machine M such that for all $y \in \Delta^*$, (i) if $y \notin L$, then Prob[M accepts y] $\leq 1/2$ and (ii) if $y \in L$, then Prob[M accepts y] $= 1$. Consider the nondeterministic polynomial time Turing machine M' constructed above and take an input $x \in \Sigma^*$. Let us compute the probability that M' accepts x. If $x \notin K$, then there exists at least one y_i ($1 \leq i \leq c$) such that $y_i \notin L$. Hence, Prob[M accepts y_i] $\leq 1/2$. With (1.8) we get Prob[M' accepts x] $\leq 1/2$. On the other hand, if $x \in K$, then $y_i \in L$ for all $1 \leq i \leq c$. Hence, Prob[M accepts y_i] $= 1$ for all $1 \leq i \leq c$. This implies that Prob[M' accepts x] $= 1$. $\square$

There is some evidence in complexity theory for **BPP** $-$ **P** and hence **RP** $=$ **coRP** $=$ **BPP** $=$ **P**. Impagliazzo and Wigderson [85] proved that if there exists a language in DTIME($2^{\mathcal{O}(n)}$) that has circuit complexity $2^{\Omega(n)}$ (which seems to be plausible), then **BPP** $=$ **P**.

An important computational problem in the context of the class **RP** is *polynomial identity testing (PIT)*. For this, we first have to define arithmetic circuits. Recall that for a ring R, $R[x_1, \ldots, x_n]$ defines the ring of polynomials with coefficients from R and variables $x_1, \ldots, x_n$. Also recall the definition of general circuits (Definition 1.32).

Definition 1.47 ((univariate, variable-free) arithmetic circuit). An *arithmetic circuit* over the ring R is a $(R[x_1, \ldots, x_n], +, \cdot, 1, -1, x_1, \ldots, x_n)$-circuit for some $n \geq 0$. Hence, an arithmetic circuit defines a polynomial from $R[x_1, \ldots, x_n]$ using the operations of addition and multiplication of polynomials, starting from the polynomials $1, -1, x_1, \ldots, x_n$. A *univariate arithmetic circuit* over the ring R is a $(R[x], +, \cdot, 1, -1, x)$-circuit. A *variable-free arithmetic circuit* over the ring R is a $(R, +, \cdot, 1, -1)$-circuit.

The characteristic char(R) of the ring R, i.e., the smallest number m such that

$$\underbrace{1 + \cdots + 1}_{m \text{ many}} = 0.$$

If no such m exists, then char(R) $= 0$. Note that for an arithmetic circuit $\mathscr{C}$ over R, all coefficients of the polynomial val($\mathscr{C}$) are from $\mathbb{Z}_{\text{char}(R)}$ (the integers modulo char(R)), where we set $\mathbb{Z}_0 = \mathbb{Z}$.

By iterated squaring, one can easily compute in logspace from a binary encoded natural number m an arithmetic circuit $\mathscr{C}$ with val($\mathscr{C}$) $= x^m$. For instance, the circuit with rhs(X_i) $= X_{i-1} X_{i-1}$ for $1 \leq i \leq n$, rhs(X_0) $= x$ and start variable X_n defines the polynomial x^{2^n}, and the size of this circuit is $n + 1$. In the same way one can define a variable-free arithmetic circuit over the ring $\mathbb{Z}$ of size $n + 1$ that defines the number 2^{2^n}. Note that the binary encoding of this number has 2^n many bits. Due to this, one may allow in arithmetic circuits right-hand sides of the form x^m for a binary encoded natural number m.

In the literature, also the term "straight-line program" is used for arithmetic circuits; see, e.g., [6, 84, 93]. Here, we reserve the term "straight-line program" for another class of circuits that will be introduced in Sect. 3.1.

The main computational problem for arithmetic circuits is PIT.

Definition 1.48 (polynomial identity testing, PIT(R)). For a ring R, the problem PIT(R) (*PIT for the ring R*) is the following computational problem:

input: An arithmetic circuit $\mathscr{C}$ over the ring R.
question: Is val($\mathscr{C}$) the zero polynomial?

Note that asking whether val($\mathscr{C}$) is the zero polynomial is not the same as asking whether val($\mathscr{C}$) is zero for all assignments of the variables in val($\mathscr{C}$): For instance, $x^2 + x \in \mathbb{Z}_2[x]$ is not the zero polynomial but $0^2 + 0 = 1^2 + 1 = 0$ in $\mathbb{Z}_2$.

It suffices to consider PIT for the rings $\mathbb{Z}$ and $\mathbb{Z}_m$ for $m \geq 2$: By the remark after Definition 1.47, PIT(R) is equivalent to PIT($\mathbb{Z}_{\text{char}(R)}$). The following important result was shown in [84] for the ring $\mathbb{Z}$ and in [3] for the rings $\mathbb{Z}_m$ ($m \geq 2$).

Theorem 1.49. *For every ring $R \in \{\mathbb{Z}\} \cup \{\mathbb{Z}_m \mid m \geq 2\}$, PIT($R$) belongs to* **coRP**.

Before we prove Theorem 1.49, let us first present a simple reduction from PIT to PIT for univariate circuits; see [3].

Proposition 1.50. *For every ring R there is a polynomial time many-one reduction from PIT to univariate PIT.*

Proof. For a polynomial $p(x_1, \ldots, x_n) \in R[x_1, \ldots, x_n]$ and a variable x_i, let $\deg(p, x_i) \in \mathbb{N}$ be the maximal number d such that $p(x_1, \ldots, x_n)$ contains a monomial of the form $r \cdot x_1^{d_1} \cdots x_{i-1}^{d_{i-1}} x_i^d x_{i+1}^{d_{i+1}} \cdots x_n^{d_n}$. Assume that $d \in \mathbb{N}$ is such that $\deg(p, x_i) < d$ for all $1 \leq i \leq n$. We define the univariate polynomial univ(p, d) $\in R[x]$ as

$$\text{univ}(p, d) = p(x^d, x^{d^2}, \ldots, x^{d^n}).$$

The mapping $p \mapsto \text{univ}(p, d)$ is injective, satisfies univ($p_1 + p_2, d$) $=$ univ(p_1, d) $+$ univ(p_2, d) and univ($p_1 \cdot p_2, d$) $=$ univ(p_1, d) $\cdot$ univ(p_2, d), and maps the zero polynomial to the zero polynomial. Hence, p is the zero polynomial if and only if univ(p, d) is the zero polynomial. Moreover, given an arithmetic circuit $\mathscr{C}$ for $p(x_1, \ldots, x_n)$, we can compute in polynomial time the binary encoding of a number d such that $\deg(p, x_i) < d$ for all $1 \leq i \leq n$. Then we obtain an arithmetic circuit for univ(p, d) by replacing every right-hand side x_i by x^{d^i}. Note that the binary encoding of the number d^i can be computed in polynomial time. □

Proof of Theorem 1.49. We prove the theorem only for the ring $\mathbb{Z}$ and $\mathbb{Z}_p$ for p prime. These are actually the only cases needed in this book. Let us start with $R = \mathbb{Z}$. Let $\mathscr{C} = (V, A_0, \text{rhs})$ be an arithmetic circuit over $\mathbb{Z}$. By Proposition 1.50 we can assume that $\mathscr{C}$ is univariate. Moreover, we can assume that $\mathscr{C}$ is in normal form. Hence, every right-hand side of $\mathscr{C}$ is either 1, -1, a variable, $X + Y$, or $X \cdot Y$ for gates X and Y. Let $p(x) = \text{val}(\mathscr{C})$ and let r be the number of gates of $\mathscr{C}$ and $k = r + 1$. The following facts follow easily by induction on r:

- The degree $\deg(p(x))$ of the polynomial $p(x)$ is at most $2^r = 2^k/2$.
- For every natural number $a \geq 2$ and every integer b with $|b| \leq a$, we have $|p(b)| \leq a^{2^r}$.

Let us define

$$A = \{1, 2, \ldots, 2^k\}.$$

Since a nonzero polynomial of degree d over a field F has at most d different roots (in our case, we can take the field $F = \mathbb{Q}$), we get that if $p(x) \neq 0$, then the probability that $p(a) = 0$ for a randomly chosen element $a \in A$ is at most $1/2$. Hence, to check whether $p(x) \neq 0$, we can guess an element $a \in A$ and check whether $p(a) \neq 0$. But here, another problem arises: the number $p(a)$ can be very large. By the above consideration, we know that $|p(a)| \leq 2^{k2^r}$ for all $a \in A$, but the binary representation of 2^{k2^r} has exponentially many bits. $\square$

To solve this problem, we compute $p(a) \bmod m$ for randomly chosen numbers m from the set

$$M = \{1, \ldots, 2^{2k}\}.$$

Claim 1. Let b be an integer with $1 \leq |b| \leq 2^{k2^k}$. Then the probability that $b \bmod m \neq 0$ for a randomly chosen number $m \in M$ is at least $1/4k$ if k is sufficiently large.

Proof of the Claim 1. By the prime number theorem, the number of primes in the set M goes to $2^{2k}/\ln(2^{2k})$ for $k \to \infty$. Hence, if k is sufficiently large, then there are more than $2^{2k}/2k$ prime numbers in M. On the other hand, if $1 \leq |b| \leq 2^{k2^k}$, then the number of different prime divisors of b is at most $\log_2(|b|) \leq k2^k$. Hence, there are at least $2^{2k}/2k - k2^k$ prime numbers $q \in M$ such that $b \bmod q \neq 0$. If k is sufficiently large, then we get

$$\frac{2^{2k}/2k - k2^k}{2^{2k}} = \frac{1}{2k} - \frac{k}{2^k} \geq \frac{1}{2k} - \frac{1}{4k} = \frac{1}{4k}.$$

This proves Claim 1. $\square$

Our algorithm works as follows:

- Randomly choose a pair $(a, m) \in A \times M$.
- Compute $p(a) \bmod m$. This can be done in polynomial time using the circuit $\mathscr{C}$ for $p(x)$. One simply replaces every gate with right-hand side x by $a \bmod m$. Then we evaluate the circuit in the ring $\mathbb{Z}_m$ of integers modulo m. Note that elements of $\mathbb{Z}_m$ can be represented with $2k = 2r + 2$ bits. Hence, all additions and multiplications can be done in polynomial time.
- If $p(a) \bmod m \neq 0$, then accept; otherwise reject.

If $p(x)$ is the zero polynomial, then this algorithm accepts with probability zero. On the other hand, if $p(x)$ is not the zero polynomial, then the algorithm accepts with probability at least $1/8k$ (if $k = r + 1$ is large enough): With probability at least $1/2$ the algorithm chooses an element $a \in A$ with $p(a) \neq 0$. Note that $p(a) \leq 2^{k2^k}$. For such an $a \in A$, the probability that the chosen element $m \in M$ satisfies $p(a) \bmod m \neq 0$ is at least $1/4k$ by Claim 1. Finally, the success probability $1/8k$ in case $p(x)$ is not the zero polynomial can be increased to $1/2$ using probability amplification. This concludes the proof for the ring $\mathbb{Z}$.

Let us now consider the field $\mathbb{F}_q$ for q a prime number. Let $\mathscr{C} = (V, A_0, \mathrm{rhs})$ be an arithmetic circuit over $\mathbb{F}_q$ in normal form and with r gates. Again, we can assume that the polynomial $\mathsf{val}(\mathscr{C}) = p(x)$ is univariate. The degree of $p(x)$ is at most 2^r. Consider the infinite field $\mathbb{F}_q(y)$ of all fractions of polynomials from $\mathbb{F}_q[y]$, which contains $\mathbb{F}_q$. Hence, we can view $p(x)$ as a polynomial with coefficients from $\mathbb{F}_q(y)$. If A is any subset of $\mathbb{F}_q[y] \subseteq \mathbb{F}_q(y)$ with $|A| \geq 2^{r+1}$, then assuming $p(x)$ is not the zero polynomial, the probability that $p(a) = 0$ for a randomly chosen polynomial $a = a(y) \in A$ is at most $1/2$. Let $d \geq 1$ be the smallest number with $q^d \geq 2^{r+1}$ (clearly, $d \leq r + 1$) and define

$$A = \{a(y) \in \mathbb{F}_q[y] \mid \deg\mathrm{ree}(a(y)) < d\}.$$

Note that $|A| = q^d \geq 2^{r+1}$. Similar to the case of characteristic 0, we will compute $p(a)$ (for $a = a(y) \in A$ chosen randomly) modulo a randomly chosen polynomial of small degree. Recall that a polynomial $c_n y^n + c_{n-1} y^{n-1} + \cdots + c_1 y + c_0$ is *monic* if $c_n = 1$. Let

$$M = \{m(y) \in \mathbb{F}_q[y] \mid m(y) \text{ is monic and } \deg\mathrm{ree}(m(y)) = 2(d + r)\}.$$

We can assume that $r \geq 1$ and hence $e := d + r \geq 2$. Note that $|M| = q^{2e}$. We need the following claim:

Claim 2. Let $b(y) \in \mathbb{F}_q[y]$ be a polynomial with $\deg\mathrm{ree}(b(y)) < 2^e$. Then the probability that $b(y) \bmod m(y) \neq 0$ for a randomly chosen polynomial $m(y) \in M$ is at least $1/8e$.

Proof of the Claim 2. The number of irreducible polynomials in M that divide $b(y)$ can be at most $2^e/2e$, because the product of all these irreducible polynomials has degree at most 2^e and every polynomial in M has degree $2e$. Since $q, e \geq 2$, we have $2^e/2e \leq q^{2e}/8e$. By [17, Eq. 3.37], the set M contains more than $q^{2e}(1 - q^{1-e})/2e \geq q^{2e}/4e$ irreducible polynomials (note that $q \geq 2$ and $e \geq 2$). Hence, the probability that $b(y) \bmod m(y) \neq 0$ for a randomly chosen polynomial $m(y) \in M$ is at least

$$\frac{q^{2e}/4e - q^{2e}/8e}{|M|} = \frac{q^{2e}/8e}{q^{2e}} = \frac{1}{8e}.$$

This proves Claim 2. □

Our probabilistic algorithm for testing whether $p(x)$ is not the zero polynomial is now very similar to the case of characteristic 0:

- Randomly choose a pair $(a(y), m(y)) \times A \times M$.
- Compute $p(a(y)) \bmod m(y)$. This can be done in polynomial time by evaluating the circuit $\mathscr{C}$ in the quotient ring $\mathbb{F}_q[y]/(m(y))$. Every element of this ring can be represented by a polynomial over $\mathbb{F}_q$ of degree at most $2(d+r) - 1 \leq 4r + 1$ and can be therefore stored in polynomial space (the prime q is a constant in our consideration). Moreover, arithmetic operations in $\mathbb{F}_q[y]/(m(y))$ can be also performed in polynomial time.
- If $p(a(y)) \bmod m(y) \neq 0$, then accept; otherwise reject.

If $p(x)$ is the zero polynomial, then this algorithm accepts with probability zero. On the other hand, if $p(x)$ is not the zero polynomial, then the algorithm accepts with probability at least $1/16(d+r)$: With probability at least $1/2$ the algorithm chooses a polynomial $a(y) \in A$ with $p(a(y)) \neq 0$. Note that the degree of $p(a(y))$ is at most $d \, 2^r < 2^{d+r}$. For such a polynomial $a(y) \in A$, the probability that the chosen polynomial $m(y) \in M$ satisfies $p(a(y)) \bmod m(y) \neq 0$ is at least $1/8(d+r)$ by Claim 2. Again, the success probability can be amplified to $1/2$. This concludes the proof for the ring $\mathbb{Z}_q$. $\qquad\square$

Let us remark that usually Theorem 1.49 is proven without a reduction to the univariate case (Proposition 1.50). In this case, one has to apply the so-called Schwartz-Zippel-DeMillo-Lipton Lemma [48, 152, 167] for multivariate polynomials. Except for this detail, the above proof follows [84].

Theorem 1.49 and the abovementioned results from [85] imply that if there exists a language in $\mathsf{DTIME}(2^{\mathcal{O}(n)})$ that has circuit complexity $2^{\Omega(n)}$, then PIT belongs to **P**. There is also an implication that goes the other way round: Kabanets and Impagliazzo [92] have shown that if PIT over the integers belongs to **P**, then one of the following conclusions holds:

- There is a language in $\mathsf{NEXPTIME} = \mathsf{NTIME}(2^{n^{\mathcal{O}(1)}})$ that does not have polynomial size circuits.
- The permanent is not computable by polynomial size arithmetic circuits.

Both conclusions represent major open problem in complexity theory. Hence, although it is quite plausible that PIT belongs to **P** (by [85]), it will be probably very hard to prove (by [92]).

We conclude this section with the following result from [6], stating that for the ring $\mathbb{Z}$, PIT is equivalent to the problem, whether a given variable- free circuit evaluates to 0.

Theorem 1.51. *There is a logspace many-one reduction from* $\mathsf{PIT}(\mathbb{Z})$ *to the following problem:*

input: A variable-free arithmetic circuit $\mathscr{C}$ over the ring $\mathbb{Z}$.
question: Is $\mathsf{val}(\mathscr{C}) = 0$?

Proof. Let $\mathscr{C}$ be an arithmetic circuit in normal form over $\mathbb{Z}$, and let $n \geq 1$ be the number of gates of $\mathscr{C}$. We construct in polynomial time a variable-free arithmetic circuit $\mathscr{C}'$ such that $\mathsf{val}(\mathscr{C}) = 0$ if and only if $\mathsf{val}(\mathscr{C}') = 0$. By Proposition 1.50, we can assume that $\mathscr{C}$ is univariate. Let $\mathsf{val}(\mathscr{C}) = p(x)$ and $a = 2^{2^{2n}} \geq 16$. By iterated squaring, one can easily construct in logspace a variable-free circuit that computes a. By plugging in this circuit into $\mathscr{C}$, we obtain a variable-free circuit $\mathscr{C}'$ that computes $p(a)$. Clearly, if $p(x) = 0$, then also $p(a) = 0$. It remains to show that $p(x) \neq 0$ implies $p(a) \neq 0$. Assume that $p(x) = \alpha x^m + \sum_{i=0}^{m-1} \alpha_i x^i$ with $\alpha \neq 0$. We have $|\alpha_i| \leq 2^{2^n}$ for $0 \leq i \leq m - 1$ and $m \leq 2^n$. To show that $p(a) \neq 0$ it suffices to show

$$\left| \sum_{i=0}^{m-1} \alpha_i a^i \right| < a^m \leq |\alpha a^m|. \tag{1.9}$$

We have

$$\left| \sum_{i=0}^{m-1} \alpha_i a^i \right| \leq 2^{2^n} \sum_{i=0}^{m-1} a^i = 2^{2^n} \cdot \frac{a^m - 1}{a - 1} \leq 2^{2^n} \cdot \frac{2a^m}{a} = a^m \cdot \frac{2^{1+2^n}}{a}$$

Hence, (1.9) follows from $a = 2^{2^{2n}} > 2^{1+2^n}$. $\qquad\qquad\square$

1.3.8 The RAM Model of Computation

Efficient algorithms are not the main concern of this work (see the textbooks [5, 42] for background on efficient algorithms). When we state that a problem belongs to **P**, we do not care about the actual polynomial: $\mathscr{O}(n)$ is as good as $\mathscr{O}(n^{100})$. Nevertheless, we make occasionally more exact statements on the running time of algorithms. In these statements we implicitly assume a more powerful computational model than Turing machines. Time $\mathscr{O}(n)$ on a Turing machine with a single work tape is very restrictive. A more realistic model that is much closer to a real computer is the *random access machine*, briefly *RAM*.

A RAM has registers $r_0, r_1, \dots$ that store integers of arbitrary size. The input is a tuple of $(i_1, \dots, i_l)$ of integers. There are elementary instructions (that all need constant time) for the following tasks:

- loading input values into registers,
- doing arithmetic operations (addition, subtraction, multiplication) on register contents,
- doing bitwise AND and OR on register contents,
- conditional jumps (if the content of a certain register is 0, then jump to the kth instruction, otherwise proceed with the next instruction),

- Indirect addressing (copy the content of register r_c into register r_i, where c is the content of register r_j). Here i and j are the arguments of the instruction.

A problem with this model is that one may produce with a sequence of $n + 1$ instructions numbers of size 2^{2^n} (which need 2^n bits): $r_0 := 2; r_0 := r_0 \cdot r_0; \ldots r_0 := r_0 \cdot r_0$, where we do n multiplications. In particular, one can evaluate a variable-free arithmetic circuit in linear time. On the other hand, in all our examples, the bit length of all registers will be linearly bounded in the input length, which for an input tuple $(i_1, \ldots, i_l)$ can be defined as $\sum_{j=1}^{l} \log |i_j|$. Under this restriction, a RAM can be easily simulated on a Turing machine with only a polynomial blowup.

Chapter 2
Preliminaries from Combinatorial Group Theory

In this chapter, we recall basic definitions from combinatorial group theory. We assume some basic knowledge of group theory; see, for instance, [149]. More background on combinatorial group theory can be found in [119, 158]. Groups will be written multiplicatively throughout this work and the group identity will be denoted with 1.

2.1 Free Groups

For an alphabet Γ let $\Gamma^{-1} = \{a^{-1} \mid a \in \Gamma\}$ be a disjoint copy of the alphabet Γ. We define $(a^{-1})^{-1} = a$, and for a word $w = a_1 a_2 \cdots a_n$ with $n \geq 0$ and $a_1, a_2, \ldots, a_n \in \Gamma \cup \Gamma^{-1}$, we define $(a_1 a_2 \cdots a_n)^{-1} = a_n^{-1} \cdots a_2^{-1} a_1^{-1}$; this defines an involution on $(\Gamma \cup \Gamma^{-1})^*$. We define the semi-Thue system

$$R_\Gamma = \{(aa^{-1}, \varepsilon) \mid a \in \Gamma \cup \Gamma^{-1}\}. \tag{2.1}$$

This system is clearly Noetherian and also locally confluent and hence confluent. The latter can be checked by considering overlapping left-hand sides. The only possible overlapping of two left-hand sides is $aa^{-1}a$ for $a \in \Gamma \cup \Gamma^{-1}$. But regardless whether we replace aa^{-1} or $a^{-1}a$ by ε, we obtain from $aa^{-1}a$ the word a. This shows local confluence.

Definition 2.1 (free group). The *free group generated by* Γ is the quotient monoid

$$F(\Gamma) = (\Gamma \cup \Gamma^{-1})^* / R_\Gamma.$$

Hence, we obtain $F(\Gamma)$ by taking the set $\mathsf{IRR}(R_\Gamma)$ of words that do not contain a factor of the form aa^{-1} for $a \in \Gamma \cup \Gamma^{-1}$. We simply call these words *irreducible* in the following. If u and v are irreducible words, then the product of u and v in the free group is the unique irreducible word that is obtained from the concatenation uv by cancelling factors of the form aa^{-1} for $a \in \Gamma \cup \Gamma^{-1}$ as long as possible.

Clearly, the identity element of $F(\Gamma)$ is the empty word ε and the inverse of the irreducible word w is w^{-1}. For $w \in (\Gamma \cup \Gamma^{-1})^*$ we denote the unique irreducible word $\mathsf{NF}_{R_\Gamma}(w)$ with $\mathsf{NF}(w)$. The word $\mathsf{NF}(w)$ is obtained from w by cancelling factors of the form aa^{-1} for $a \in \Gamma \cup \Gamma^{-1}$ as long as possible in any order. For example, we have $\mathsf{NF}(a^{-1}aa^{-1}bb^{-1}b) = a^{-1}b$.

2.2　Finitely Generated Groups and Finitely Presented Groups

Definition 2.2 (finitely generated group). A group G is *finitely generated* if there exists a finite set $A \subseteq G$ such that every $g \in G$ can be written as a product $a_1 a_2 \cdots a_n$ with $a_1, a_2, \ldots, a_n \in A \cup A^{-1}$. The set A is also called a *finite generating set* for G.

Equivalently, G is finitely generated if there exists a finite set Γ and a surjective group homomorphism $h : F(\Gamma) \to G$.

For a subset $B \subseteq G$ we denote with $\langle B \rangle$ the subgroup of G generated by B; it contains all products $b_1 b_2 \cdots b_n$ with $b_1, b_2, \ldots, b_n \in B \cup B^{-1}$. The *normal closure of* $B \subseteq G$, denoted by $\langle B \rangle^G$ is the subgroup generated by all conjugates of elements from B, i.e.,

$$\langle B \rangle^G = \langle \{ gbg^{-1} \mid g \in G, b \in B \} \rangle.$$

Following common notation, the quotient group $G / \langle B \rangle^G$ is also denoted with $\langle G \mid B \rangle$.

Definition 2.3 (presentation of a group, relator, finitely presented group). For a set Γ and $R \subseteq F(\Gamma)$ we write $\langle \Gamma \mid R \rangle$ for the group $\langle F(\Gamma) \mid R \rangle$. The set R is called a set of *relators* of the group $\langle \Gamma \mid R \rangle$, and the pair (Γ, R) is a *presentation* of the group $\langle G \mid R \rangle$. Here, Γ and R can be infinite. If $\Gamma = \{ a_1, \ldots, a_n \}$ and $R = \{ r_1, \ldots, r_m \}$ are finite, we also write $\langle a_1, \ldots, a_n \mid r_1, \ldots, r_m \rangle$ for $\langle \Gamma \mid R \rangle$. If G is finitely generated by $\{ a_1, \ldots, a_n \}$, then G can be written in the form $\langle a_1, \ldots, a_n \mid R \rangle$ for a set $R \subseteq F(\Gamma)$, which in general is infinite. A group G is *finitely presented* if G is isomorphic to a group $\langle a_1, \ldots, a_n \mid r_1, \ldots, r_m \rangle$.

Example 2.4. Consider the presentation $(\{a, b\}, \{[a, b], a^4, b^2\})$. Here,

$$[a, b] = aba^{-1}b^{-1}$$

is the *commutator* of a and b. The relator $[a, b]$ specifies that a and b commute. Moreover, a (resp., b) generates a copy of the cyclic group $\mathbb{Z}_4$ (resp., $\mathbb{Z}_2$). Hence, we have $\langle a, b, \mid [a, b], a^4, b^2 \rangle \cong \mathbb{Z}_4 \times \mathbb{Z}_2$.

Let us introduce some important classes of finitely presented groups that will reoccur later.

Example 2.5 (graph groups). Let (A, I) be a finite undirected graph, where $I \subseteq A \times A$ is the irreflexive and symmetric edge relation. With (A, I) we associate the *graph group*

$$G(A, I) = \langle A \mid [a, b] \text{ for } (a, b) \in I \rangle.$$

In other words, two generators are allowed to commute if and only if they are adjacent in the graph. Graph groups are also known as *right-angled Artin groups* or *free partially commutative groups*. For a survey on graph groups see [32].

Example 2.6 (Coxeter groups). Let $M = (m_{i,j})_{1 \leq i, j \leq n}$ be a symmetric $(n \times n)$-matrix over $\mathbb{N}$ such that $m_{i,j} = 1$ if and only if $i = j$. The corresponding *Coxeter group* is

$$C(M) = \langle a_1, \ldots, a_n \mid (a_i a_j)^{m_{i,j}} \text{ for } 1 \leq i, j \leq n \rangle.$$

In particular, $a_i^2 = 1$ in $C(M)$ for $1 \leq i \leq n$. Traditionally, one writes the entry ∞ instead of 0 in the *Coxeter matrix* M, and then $m_{i,j}$ becomes the order of the element $a_i a_j$. If $m_{i,j} \in \{0, 1, 2\}$ for all $1 \leq i, j \leq n$, then $C(M)$ is called a *right-angled Coxeter group*. In this case we can define $C(M)$ also as $\langle G(A, I) \mid a_i^2 \text{ for } 1 \leq i \leq n \rangle$, where (A, I) is the graph with $A = \{a_1, \ldots, a_n\}$ and $(a_i, a_j) \in I$ if and only if $m_{i,j} = 2$. The book [20] provides a detailed introduction into Coxeter groups.

Example 2.7 (automatic groups). To define automatic groups, let us first define the convolution of words. For an alphabet A and two words $u = a_1 a_2 \cdots a_m$ and $v = b_1 b_2 \cdots b_n$ (with $a_1, \ldots, a_m, b_1, \ldots, b_n \in A$), the *convolution* $u \otimes v$ is the following word over the alphabet $(A \cup \{\#\}) \times (A \cup \{\#\}) \setminus \{(\#, \#)\}$, where $\# \notin A$, $k = \max\{m, n\}$, $a_i = \#$ for $m < i \leq k$, and $b_i = \#$ for $n < i \leq k$:

$$u \otimes v = (a_1, b_1)(a_2, b_2) \cdots (a_k, b_k).$$

For a finitely generated group G, an *automatic structure* for G consists of a triple $(\Gamma, A, (A_a)_{a \in \Gamma \cup \{1\}})$, where Γ is a finite generating set for G, A is a finite automaton over $\Gamma \cup \Gamma^{-1}$ such that the canonical homomorphism $h : F(\Gamma) \to G$ is surjective on $L(A)$ [where $L(A)$ is viewed as a subset of $F(\Gamma)$], and for every $a \in \Gamma \cup \{1\}$, A_a is a finite automaton over the alphabet $(\Gamma \cup \Gamma^{-1} \cup \{\#\}) \times (\Gamma \cup \Gamma^{-1} \cup \{\#\}) \setminus \{(\#, \#)\}$ such that for all $u, v \in L(A)$, $u \otimes v \in L(A_a)$ if and only if $h(ua) = h(v)$. Then, a group G is *automatic* if it has an automatic structure. The book [57] provides a detailed introduction into automatic groups. Every automatic group is finitely presented. Graph groups as well as Coxeter groups are automatic; see [24, 30, 75], respectively.

Sometimes, we will write an identity $u = v$ instead of the relator uv^{-1} in a group presentation. For instance, the graph group $G(A, I) = \langle A \mid [a, b]$ for $(a, b) \in I \rangle$ could be also written as $\langle A \mid ab = ba$ for $(a, b) \in I \rangle$. Moreover, for words $u, v \in (\Gamma \cup \Gamma^{-1})^*$ we say that "$u = v$ in $\langle \Gamma \mid R \rangle$" if $uv^{-1} \in \langle R \rangle^{F(\Gamma)}$. This means that u and v represent the same group element of $\langle \Gamma \mid R \rangle$. We can also obtain the group $\langle \Gamma \mid R \rangle$ as a quotient monoid by taking the semi-Thue system $S = R_\Gamma \cup \{(r, \varepsilon) \mid r \in R \cup R^{-1}\}$, where R_Γ was defined in (2.1). Then

$$\langle \Gamma \mid R \rangle \cong (\Gamma \cup \Gamma^{-1})^* / S.$$

Clearly, every group G can be written in the form $\langle \Gamma \mid R \rangle$. Simply take $\Gamma = G$ and R as the kernel of the canonical homomorphism $h : F(G) \to G$. The notation $\langle \Gamma \mid R \rangle$ allows to define the free product of two groups in the following way.

Definition 2.8 (free product). Let $G_1 = \langle \Gamma_1 \mid R_1 \rangle$ and $G_1 = \langle \Gamma_2 \mid R_2 \rangle$ be groups with $\Gamma_1 \cap \Gamma_2 = \emptyset$. Then the *free product* $G_1 * G_2$ is $\langle \Gamma_1 \cup \Gamma_2 \mid R_1 \cup R_2 \rangle$.

One can think about the free product $G_1 * G_2$ (where we assume $G_1 \cap G_2 = \{1\}$) as the set of all alternating sequences $g_1 g_2 \ldots g_n$, where every g_i belongs to either $G_1 \setminus \{1\}$ or $G_2 \setminus \{1\}$ and $g_i \in G_1 \setminus \{1\}$ if and only if $g_{i+1} \in G_2 \setminus \{1\}$ for all $1 \le i \le n-1$. The product of two such sequences is computed in the natural way by multiplying elements from the same group (G_1 or G_2) as long as possible. Formally, let $g = g_1 g_2 \cdots g_m$ and $h = h_1 h_2 \cdots h_n$ be alternating sequences as described above. If $m = 0$, then $gh = h$, and if $n = 0$, then $gh = g$. Now, assume that $n, m > 0$. If g_m and h_1 are from different groups, i.e., $g_m \in G_1 \setminus \{1\}$ if and only if $h_1 \in G_2 \setminus \{1\}$, then gh is the alternating sequence $g_1 g_2 \cdots g_m h_1 h_2 \cdots h_n$. Finally, assume that g_m and h_1 are from the same group. Let k be the maximal index $0 \le k \le \min\{n, m\}$ such that for all $1 \le i \le k$, $g_{m-i+1} = h_i^{-1}$. If $k = m$, then $gh = h_{i+1} \cdots h_n$, and if $k = n$, then $gh = g_1 \cdots g_{m-k}$ (this includes the case $k = m = n$ with $gh = 1$). Finally, if $k < \min\{m, n\}$, then $gh = g_1 \cdots g_{m-k-1}(g_{m-k}h_{i+1})h_{i+2} \cdots h_n$.

If G is a group, and R is a subset of $G * F(\Gamma)$, then we write $\langle G, \Gamma \mid R \rangle$ (or $\langle G, a_1, \ldots, a_n \mid R \rangle$ if $\Gamma = \{a_1, \ldots, a_n\}$) for the group $\langle G * F(\Gamma) \mid R \rangle$.

2.3 Tietze Transformations

Already in 1908, Tietze came up with a semi-decision procedure for checking whether two group presentations yield isomorphic groups [160]. His procedure consists of four elementary transformation steps.

Definition 2.9 (Tietze transformations). Let (Γ, R) be a finite presentation of a group:

- If $r \in \langle R \rangle^{F(\Gamma)}$, then we say that the presentation $(\Gamma, R \cup \{r\})$ is obtained from a *Tietze-type-1 transformation* from (Γ, R). Intuitively, we add a redundant relator to the presentation.
- If $r \in R$ is such that $r \in \langle R \setminus \{r\} \rangle^{F(\Gamma)}$, then we say that the presentation $(\Gamma, R \setminus \{r\})$ is obtained from a *Tietze-type-2 transformation* from (Γ, R). Intuitively, we remove a redundant relator from the presentation.
- If $a \notin \Gamma \cup \Gamma^{-1}$ and $w \in (\Gamma \cup \Gamma^{-1})^*$, then we say that the presentation $(\Gamma \cup \{a\}, R \cup \{a^{-1}w\})$ is obtained from a *Tietze-type-3 transformation* from (Γ, R). Intuitively, we add a redundant generator to the presentation. Note that the new relator $a^{-1}w$ defines the new generator a as w (a word over the old generators).
- If $a \in \Gamma$ and $w \in ((\Gamma \cup \Gamma^{-1}) \setminus \{a, a^{-1}\})^*$ such that $a^{-1}w \in R$, then we say that the presentation $(\Gamma \setminus \{a\}, \varphi(R))$ is obtained from a *Tietze-type-4 transformation* from (Γ, R). Here, φ is the homomorphism with $\varphi(a) = w$, $\varphi(a^{-1}) = w^{-1}$, and $\varphi(c) = c$ for $c \in (\Gamma \cup \Gamma^{-1}) \setminus \{a, a^{-1}\}$. Intuitively, $a = w$ holds in the group $\langle \Gamma \mid R \rangle$; hence, a is redundant and can be removed from the presentation.

Any of the above four transformations is called a *Tietze transformation*.

It is not hard to see that if the finite group presentation (Σ, S) is obtained by a finite sequence of Tietze transformations from (Γ, R), then $\langle \Gamma \mid R \rangle \cong \langle \Sigma \mid S \rangle$. Tietze [160] proved that also the converse holds.

Theorem 2.10. *Let (Γ, R) and (Σ, S) be finite group presentations. Then $\langle \Gamma \mid R \rangle \cong \langle \Sigma \mid S \rangle$ if and only if (Σ, S) can be obtained by a finite sequence of Tietze transformations from (Γ, R).*

This result implies that the isomorphism problem for finitely generated groups (i.e., the question, whether two given finite group presentations yield isomorphic groups) is computably enumerable. We will use Tietze transformations in Chap. 6.

Example 2.11. Consider the group $\langle a, b \mid [a, b], a^3, b^2 \rangle \cong \mathbb{Z}_3 \times \mathbb{Z}_2 \cong \mathbb{Z}_6 \cong \langle c \mid c^6 \rangle$. Hence, we should be able to derive the presentation $(\{c\}, \{c^6\})$ from the presentation $(\{a, b\}, \{[a, b], a^3, b^2\})$ using Tietze transformations. Here is a derivation:

1. A Tietze-type-3 transformation allows to go from $(\{a, b\}, \{[a, b], a^3, b^2\})$ to $(\{a, b, c\}, \{[a, b], a^3, b^2, c^{-1}ab\})$. In the following, we only list the current set of relators. Moreover, we write the relators $[a, b], a^3, b^2, c^{-1}ab$ in the more readable form $ab = ba, a^3 = 1, b^2 = 1, c = ab$.
2. Using a Tietze-type-1 transformation (adding the consequence $c = ba$) followed by a Tietze-type-2 transformation (removing the redundant relator $ab = ba$), we get $c = ba, a^3 = 1, b^2 = 1, c = ab$.
3. The relator $c = ab$ can be written as $a = cb^{-1}$. This allows to eliminate the generator a with a Tietze-type-4 transformation. We get $c = bcb^{-1}, (cb^{-1})^3 = 1, b^2 = 1$.
4. Since $b = b^{-1}$ follows from $b^2 = 1$, we get $c = bcb, (cb)^3 = 1, b^2 = 1$ using a Tietze-type-1 and Tietze-type-2 transformations.

5. Next, $cbcbcb = 1$ can be replaced by $c^3 b = 1$ (since $bcb = c$). We now have
   $c = bcb$, $c^3 b = 1$, $b^2 = 1$, or, equivalently, $c = bcb$, $c^3 = b$, $b^2 = 1$.
6. A Tietze-type-4 transformation (eliminating b) yields $c = c^7$, $c^3 = 1$.
7. Finally, a Tietze-type-2 transformation allows to remove the relator $c = c^7$, and
   we arrive at the presentation $(\{c\}, \{c^6\})$.

2.4 The Word Problem

Now that we have defined basic concepts from combinatorial group theory and complexity theory, it is time to introduce the word problem, which is the fundamental computational problem in combinatorial group theory. Basically, the word problem for a finitely generated group G asks whether a given word over the generators of G and their inverses represents the group identity. Here is the formal definition.

Definition 2.12. Let G be a finitely generated group with a finite generating set Γ; hence G is isomorphic to $\langle \Gamma \mid R \rangle$ for a set of relators $R \subseteq (\Gamma \cup \Gamma^{-1})^*$. The *word problem for G with respect to Γ*, briefly $\mathsf{WP}(G, \Gamma)$, is the following problem:

input: A word $w \in (\Gamma \cup \Gamma^{-1})$.
question: Does $w = 1$ hold in G (i.e., does w belong to $\langle R \rangle^{F(\Gamma)}$)?

In the way we defined it, the word problem depends on the chosen generating set for G. But if we are only interested in the decidability/complexity of word problems, the actual generating set is not relevant.

Lemma 2.13. *Let G be a finitely generated group and let Γ and Σ be two finite generating sets. Then $\mathsf{WP}(G, \Gamma) \leq_m^{\log} \mathsf{WP}(G, \Sigma)$.*

Proof. Let $G \cong \langle \Gamma \mid R \rangle \cong \langle \Sigma \mid S \rangle$. There exists a homomorphism $h : (\Gamma \cup \Gamma^{-1})^* \to (\Sigma \cup \Sigma^{-1})^*$ with $h(a^{-1}) = h(a)^{-1}$ for all $a \in \Gamma \cup \Gamma^{-1}$ that induces an isomorphism from $\langle \Gamma \mid R \rangle$ to $\langle \Sigma \mid S \rangle$. Hence, for a word $w \in (\Gamma \cup \Gamma^{-1})^*$ we have $w = 1$ in $\langle \Gamma \mid R \rangle$ if and only if $h(w) = 1$ in $\langle \Sigma \mid S \rangle$. The lemma follows, since the homomorphism h can be computed with a logspace transducer; see Example 1.25.

□

Because of Lemma 2.13, it is justified to just speak of the word problem for G, briefly $\mathsf{WP}(G)$. It is not difficult to come up with a finitely generated group with an undecidable word problem. For finitely presented groups, this is a much harder problem that was finally solved by Novikov [136] and independently by Boone [22] in the 1950s.

Theorem 2.14. *There is a finitely presented group with an undecidable word problem.*

A modern treatment of this result can be found [158]. In this context, we should also mention the following celebrated result by Higman that is known as Higman's embedding theorem [77].

Theorem 2.15. *Let G be a finitely generated group. Then G has a computably enumerable word problem if and only if G can be embedded into a finitely presented group.*

An algebraic characterization of groups with a computable word problem exists as well; it goes back to Boone and Higman [23].

Theorem 2.16. *Let G be a finitely generated group. Then G has a computable word problem if and only if G can be embedded into a simple subgroup of a finitely presented group.*

In this book, we are only interested in groups with a computable word problem. The class of groups with a computable word problem contains many important group classes. Finitely presented groups with undecidable word problem are somehow exotic. The relators of such a group encode the transition relation of a Turing machine with an undecidable halting problem.

If a group G has a computable word problem, the next question addresses the complexity of the word problem. Actually, for many classes of groups, the word problem can be solved in polynomial time. Here are some examples:

- Word hyperbolic groups [68]: Dehn's algorithm (see [119, Chap. V.4]) solves the word problem for these groups in linear time. Actually, the word problem for a hyperbolic group can be solved in real time (i.e., the time bound is the length of the input word) on a Turing machine with several work tapes [81]. In [25], it was shown that the word problem for a hyperbolic group belongs to the circuit complexity class $\mathbf{NC}^2$. In [111] this upper bound was slightly improved to $\mathbf{LogCFL}$, which is the class of all languages that are logspace many-one reducible to a context-free language. Actually, the argument in [111] shows that every group having a Cannon's algorithm (a generalization of Dehn's algorithm) [64] has a word problem in $\mathbf{LogCFL}$. Besides word hyperbolic groups, every finitely generated nilpotent group has a Cannon's algorithm [64].
- Graph groups; see Example 2.5: By [49,166] the word problem for a graph group can be solved in linear time on a RAM.
- Automatic groups; see Example 2.7: The word problem can be solved in quadratic time on a Turing machine.

Concerning space complexity, there is actually a big class of groups, where the word problem can be even solved in logarithmic space.

Definition 2.17. A group G is *linear* if there exist a field F and a dimension d such that G embeds into the *general linear group* $\mathsf{GL}_d(F)$ (the group of all invertible $(d \times d)$-matrices over the field F).

The following result is shown in [108] for a field F of characteristic 0 and in [155] for a field of prime characteristic.

Theorem 2.18. *Let G be a finitely generated linear group. Then the word problem for G belongs to $\mathbf{L}$.*

The proof of Theorem 2.18 makes essential use of the following result, which we will use in Sect. 4.5. Implicitly it appears in [108] for the case of a field of characteristic 0; see also [155]. Recall that for a field F, $F(x_1, \ldots, x_m)$ denotes the field of all fractions of polynomials from $F[x_1, \ldots, x_m]$.

Theorem 2.19. *Let G be a finitely generated linear group over a field F and let P be the prime field of F (which is either $\mathbb{Q}$ if F has characteristic 0 or $\mathbb{F}_p$ if F has prime characteristic p). Then, G is isomorphic to a group of matrices over the field $P(x_1, \ldots, x_m)$.*

Proof. The proof uses a bit of field theory. Let G be a finitely generated subgroup of $\mathsf{GL}_n(F)$. Hence, there exists a finite subset $A \subseteq F$ such that every matrix entry from a generator of G belongs to A. We can therefore replace the field F by the subfield K generated by A. If we choose a maximal algebraically independent subset $\{x_1, \ldots, x_m\} \subseteq A$, then K becomes isomorphic to a finite algebraic extension of a field of fractions $P(x_1, \ldots, x_m)$, where P is the prime field of F; see, e.g., [86, p. 156]. Let $[K : P(x_1, \ldots, x_m)] = d$ be the degree of this algebraic extension. The field K can be also seen as a d-dimensional associative algebra over the base field $P(x_1, \ldots, x_m)$ (i.e., a d-dimensional vector space over $P(x_1, \ldots, x_m)$ with a multiplication that together with the vector addition yields a ring structure). By the regular representation for associative algebras [87], K is isomorphic to a subring of the ring $M_d(P(x_1, \ldots, x_m))$ of $(d \times d)$-matrices over $P(x_1, \ldots, x_m)$. Since K is a field, every nonzero element of K is thereby represented by a matrix from $\mathsf{GL}_d(P(x_1, \ldots, x_m))$. Hence, by replacing in the generator matrices of G every matrix entry by a $(d \times d)$-matrix over $P(x_1, \ldots, x_m)$, it follows that G is isomorphic to a finitely generated subgroup of $\mathsf{GL}_{nd}(P(x_1, \ldots, x_m))$.   □

Let us briefly sketch how Theorem 2.18 for characteristic 0 can be deduced from Theorem 2.19. By Theorem 2.19 it remains to show that an identity $A_1 A_2 \cdots A_n = \mathsf{Id}_k$ can be tested in logspace, where $A_1, A_2, \ldots, A_n$ are from a fixed set of $(k \times k)$-matrices over $\mathbb{Q}(x_1, \ldots, x_m)$. This identity can be reduced to an identity $B_1 B_2 \cdots B_n = C_1 C_2 \cdots C_n$, where $B_1, C_1, \ldots, B_n, C_n$ are from a fixed set of $(k \times k)$-matrices over $\mathbb{Z}[x_1, \ldots, x_m]$. Let d be the maximal degree of a polynomial in this set. Hence, in the product matrices $B_1 B_2 \cdots B_n$ and $C_1 C_2 \cdots C_n$ every polynomial has degree at most dn. It follows that $B_1 B_2 \cdots B_n = C_1 C_2 \cdots C_n$ if and only if for every m-tuple $\overline{a} \in \{0, \ldots, dn\}^m$, we have $B_1(\overline{a}) B_2(\overline{a}) \cdots B_n(\overline{a}) = C_1(\overline{a}) C_2(\overline{a}) \cdots C_n(\overline{a})$ (we denote with $B_i(\overline{a})$ the integer matrix obtained by replacing every polynomial entry $p(x_1, \ldots, x_m)$ of B_i by $p(\overline{a})$ and similarly for the matrices C_i. Since m is a constant in our consideration, we can consider every tuple $\overline{a} \in \{0, \ldots, dn\}^m$ one after the other and thereby reuse the space needed for the previous tuple. Hence, it remains to verify in logspace an identity $B_1 B_2 \cdots B_n = C_1 C_2 \cdots C_n$, where $B_1, C_1, \ldots, B_n, C_n$ are $(k \times k)$-matrices over $\mathbb{Z}$ with entries bounded in absolute value by $n^{\mathcal{O}(1)}$. One can easily show that every entry in the product matrices $B_1 B_2 \cdots B_n$ and $C_1 C_2 \cdots C_n$ has size $n^{\mathcal{O}(n)}$. Using the Chinese remainder theorem, it suffices to verify the identity modulo of the first ℓ prime numbers, where $\ell \in n^{\mathcal{O}(1)}$. Actually, we can check the identity modulo of all numbers $2, \ldots, p$, where p is the ℓth prime number. Note that p has only $\mathcal{O}(\log n)$ many bits. Hence, all computations can be done in logspace.

Examples of finitely generated linear groups are finitely generated polycyclic groups, Coxeter groups, braid groups, and graph groups. Hence, for all these groups the word problem can be solved in logspace.

Finitely generated *metabelian group* (a group G is metabelian if it is has a normal abelian subgroup A such that the quotient G/A is abelian too) can be embedded into finite direct products of linear groups [164]. Hence, also for finitely generated metabelian groups the word problem can be solved in logspace. An interesting class of groups, for which in general no efficient algorithm for the word problem is known, is the class of *one-relator groups*, i.e., groups of the form $\langle \Gamma \mid r \rangle$ for $r \in (\Gamma \cup \Gamma^{-1})^*$. By a famous result of Mangus [121] every one-relator group has a computable word problem, which moreover is primitive recursive [10, 29]. No better general complexity bound is known. On the other hand, no example of a one-relator group with a provably hard word problem is known. Let us also mention that by recent results of Wise [165], every one-relator group with torsion is linear and hence has a logspace word problem by Theorem 2.18.

By adapting the Novikov-Boone construction, Cannonito and Gatterdam proved that for every $n \geq 0$, there is a finitely presented group G such that (i) WP(G) is computable but (ii) not within the nth level of the Grzegorczyk hierarchy [27]. Recently, it was even shown that for every computable set A there is a finitely presented *residually finite group* G (recall that G is residually finite if for every $g \in G \setminus 1$ there exists a homomorphism $h : G \to H$ with H finite and $h(g) \neq 1$) such that WP(G) is at least as hard as A [97]. McKinsey [126] has shown that every finitely presented residually finite group has a computable word problem.

2.5 The Dehn Function and the Word Search Problem

In this section, we will introduce the Dehn function of a group, whose growth is an important geometric invariant of a finitely presented group. A beautiful survey on Dehn functions can be found in [147]. Finally, we will briefly discuss the word search problem for a finitely presented group. This is an extension of the word problem, where we not only ask whether a given word w represents the identity 1, but, in the positive case, also want to compute a witness (in form of a van Kampen diagram) for the fact that $w = 1$.

Assume that we have a finitely presented group $G = \langle \Gamma \mid R \rangle$ where $R \subseteq (\Gamma \cup \Gamma^{-1})^+$ is a finite set of cyclically reduced words of length at least 2 (an irreducible word w is cyclically reduced if w is of the form avb for $b \neq a^{-1}$). We also say that the finite presentation (Γ, R) is *reduced*, and it is easy to see that a given finite presentation for G can be transformed (in polynomial time) into a reduced finite

Fig. 2.1 The starting point
for folding

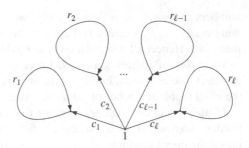

presentation for G. If an irreducible word $w \in (\Gamma \cup \Gamma^{-1})^*$ represents the identity of G, then in the free group $F(\Gamma)$, w belongs to the normal subgroup $\langle R \rangle^{F(\Gamma)}$. Hence, in $F(\Gamma)$ we can write w as a product $\prod_{i=1}^{\ell} c_i r_i c_i^{-1}$ with $c_i \in (\Gamma \cup \Gamma^{-1})^*$ irreducible and $r_i \in R \cup R^{-1}$. The smallest number ℓ for which we can write w as such a product is also called the *area* of w [with respect to the presentation (Γ, R)]. The choice of the term "area" can be explained as follows. Given the word $\prod_{i=1}^{\ell} c_i r_i c_i^{-1}$ one obtains a planar diagram as follows: start with a (planar) bouquet as shown in Fig. 2.1. Here, the c_i-labeled edge stands for a path of $(\Gamma \cup \Gamma^{-1})$-labeled edges such that the concatenation of the labels yields c_i and similarly for the r_i-labeled loop. Moreover, for every a-labeled edge there is a reversed a^{-1}-labeled edge. If we walk along the boundary of this bouquet in clockwise orientation, starting in the base point 1 (this is the so-called boundary cycle), then we read exactly the word $\prod_{i=1}^{\ell} c_i r_i c_i^{-1}$. The reduction of this word using the rewriting system R_Γ for the free group $F(\Gamma)$ [see (2.1)] yields the irreducible word w. This reduction process on words translates into a folding process on planar diagrams. As long as the boundary cycle contains two successive edges e and f, such that the label of f is the inverse of the label of e, then we fold these two edges into one edge and thereby identify the initial vertex of e with the target vertex of f. In case the initial vertex of e is already equal to the target vertex of f (this situation can arise, for instance, when folding the bouquet for rr^{-1} for a relator r), then e and f enclose a region, and we remove this whole region including all edges and nodes within the region. This is a confluent process, i.e., the completely folded planar diagram obtained this way does not depend from the chosen order of folding steps. Moreover, the boundary cycle of the completely folded planar diagram is labeled with the word w. Such a diagram is called a *van Kampen diagram* for w and its area is the number of cells (closed regions) [119]. Hence, the area of an irreducible word w with $w = 1$ in G is the minimal area of a van Kampen diagram for w. The boundary of any cell of a van Kampen diagram is labeled with a relator or the inverse of a relator.

Example 2.20. Let us consider the presentation $(\{a, b\}, \{aba^{-1}b^{-1}\})$ of $\mathbb{Z} \times \mathbb{Z}$. Let us fold the bouquet corresponding to

$$w = a^{-1}(aba^{-1}b^{-1})a\ ba^{-1}(aba^{-1}b^{-1})ab^{-1}$$

which is (we do not show the inverse edges labeled with a^{-1} or b^{-1}):

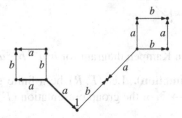

The origin is labeled with 1 and the two thick edges are the one's that are folded first. Hence, we get:

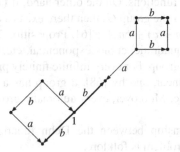

Folding the two b-labeled edges at the origin 1 yields the following diagram:

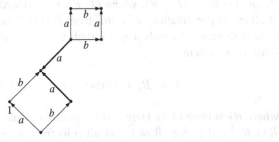

Two more folding steps yield the following diagrams:

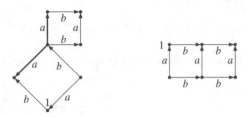

The right diagram is a van Kampen diagram for $b^2a^{-1}b^{-2}a$ of area 2.

Definition 2.21 (Dehn function). Let (Γ, R) be a finite group presentation. The *Dehn function* $D_{\Gamma,R} : \mathbb{N} \to \mathbb{N}$ of the group presentation (Γ, R) is defined as

$$D_{\Gamma,R}(n) = \max\{\text{area}(w) \mid w \text{ is irreducible}, w \in \langle R \rangle^{F(\Gamma)}, |w| \leq n\}.$$

In other words, $D_{\Gamma,R}(n)$ is the maximal area of an irreducible word of length at most n that represents the identity of G. Clearly, different finite presentations for G may yield different Dehn functions. On the other hand, if (Γ_1, R_1) and (Γ_2, R_2) are two finite presentations for the group G, then there exists a constant $c > 0$ such that $D_{\Gamma_1,R_1}(n) \leq c \cdot D_{\Gamma_2,R_2}(cn + c) + cn + c$ [61, Proposition 2.4]. Hence, whether the Dehn function is linear, quadratic, cubic, exponential, etc. does not depend on the chosen presentation for a group. For every infinite finitely presented group the Dehn function grows at least linear, and by [68], a group has a linear Dehn function if and only if it is hyperbolic. Moreover, every automatic group has a quadratic Dehn function.

The following relationship between the Dehn function of a group and the complexity of the word problem is folklore.

Proposition 2.22. *Let $D(n)$ be a Dehn function for the finitely presented group G. Then the word problem for the group G can be solved in* NTIME$(\mathcal{O}(D(n)))$.

Proof. Let $G \cong \langle \Gamma \mid R \rangle$, where (Γ, R) is a reduced presentation. Let S be the set of all cyclic permutations of words from $R \cup R^{-1}$ (a word s is a cyclic permutation of w if there exist words u and v such that $w = uv$ and $s = vu$). We define the semi-Thue system

$$T = R_\Gamma \cup \{(u, v) \mid uv^{-1} \in R \cup R^{-1}, u \neq \varepsilon\},$$

where R_Γ is from (2.1). Here is some intuition: Rules of the form (u, v) with $uv^{-1} \in R \cup R^{-1}$ and $u \neq \varepsilon$ allow to cut off cells from a van Kampen diagram that have a nontrivial intersection with the boundary, where u is the part of the cell that belongs to the boundary. Rules (aa^{-1}, ε) allow to remove boundary edges e, where one endpoint is only adjacent with edge e. Using this intuition, the following fact is not hard to see; see, e.g., [147]. For a given irreducible input word w of length n, $w = 1$

in $\langle \Gamma \mid R \rangle$ if and only if w can be rewritten with T to the empty word in at most $(m+1)D_{\Gamma,R}(n)+n$ steps, where m the maximal length of a relator in R (which is a constant in our consideration). A nondeterministic Turing machine simply guesses such a sequence of rewrite steps. Moreover, successive rewrite steps can be applied within bounded distance in the following sense: If the word xuv is rewritten to xvy using the rule $(u, v) \in T$, then (i) either $v = \varepsilon$ and the next rewritten factor covers the last symbol of x or the first symbol of y or (ii) $v \neq \varepsilon$ and the next rewritten factor covers a position in v. Hence, the whole reduction process can be carried out in time $\mathcal{O}(D_{\Gamma,R}(n) + n)$ on a Turing machine. The proposition follows, since $D_{\Gamma,R}(n)$ grows at least linear if $\langle \Gamma \mid R \rangle$ is infinite. $\qquad \square$

Another important and much more difficult result in this context is the following **NP**-analogue of Higman's embedding theorem, which was shown by Birget, Ol'shanskii, and Rips [19].

Theorem 2.23. *For a finitely generated group G, the word problem belongs to* **NP** *if and only if G can be embedded into a finitely presented group with a polynomial Dehn function.*

The Dehn function for a finitely presented group G only yields an upper for the complexity of the word problem. For instance, the Baumslag-Solitar group $\langle a, b \mid bab^{-1} = a^2 \rangle$ has an exponential Dehn function [57], but can be embedded into the linear group $\mathsf{GL}_2(\mathbb{Q})$ using

$$a \mapsto \begin{pmatrix} 1 & 1 \\ 0 & 1 \end{pmatrix}, \quad b \mapsto \begin{pmatrix} 1/2 & 0 \\ 0 & 1 \end{pmatrix}.$$

Hence, by Theorem 2.18 the word problem for $\langle a, b \mid bab^{-1} = a^2 \rangle$ can be solved in logarithmic space. An even more extreme example is the Baumslag-Gersten group $\mathrm{BG}(1, 2) = \langle a, b, t \mid bab^{-1} = t, tat^{-1} = a^2 \rangle$, which was introduced by Baumslag [13] as the first example of a noncyclic group, all of whose finite quotients are cyclic. Its Dehn function is roughly a tower of $\log_2(n)$ exponents and hence nonelementary [61, 143]. Nevertheless, using power circuits [133], which are a succinct representation of huge integers, the word problem for $\mathrm{BG}(1, 2)$ can be solved in polynomial time [54, 132]. In [38] examples of groups with a computable word problem are constructed, for which the gap between the complexity of the word problem and the Dehn function is not bounded by a primitive recursive function. In [97] this result is further improved in two aspects: The gap is increased to any recursive function and the group that realizes the gap is finitely presented and residually finite.

Definition 2.24. Let (Γ, R) be a finite reduced presentation and let $G=\langle \Gamma \mid R \rangle$. The *word search problem* for (Γ, R), briefly $\mathsf{WSP}(\Gamma, R)$, is the following computational problem with output:

input: An irreducible word $w \in (\Gamma \cup \Gamma^{-1})^*$.
output: "No" if $w \neq 1$ in G; otherwise irreducible words $c_1, \ldots, c_\ell \in (\Gamma \cup \Gamma^{-1})^*$
  and $r_1, \ldots, r_\ell \in R \cup R^{-1}$ such that $w = \prod_{i=1}^{\ell} c_i r_i c_i^{-1}$ in $F(\Gamma)$.

Hence, instead of just asking whether a given word represents the identity of G, one also wants to have a proof for representing the identity in the positive case. Instead of returning irreducible words $c_1, \ldots, c_\ell \in (\Gamma \cup \Gamma^{-1})^*$ and $r_1, \ldots, r_\ell \in R \cup R^{-1}$ such that $w = \prod_{i=1}^{\ell} c_i r_i c_i^{-1}$ in $F(\Gamma)$, one might also return a van Kampen diagram with boundary w. Both variants are equivalent with respect to polynomial time computations: From the word $\prod_{i=1}^{\ell} c_i r_i c_i^{-1}$ one can compute in polynomial time a van Kampen diagram with boundary w, and from a van Kampen diagram with boundary w, one can compute in polynomial time a word $\prod_{i=1}^{\ell} c_i r_i c_i^{-1}$ that is equal to w in the free group $F(\Gamma)$.

Logspace many-one reductions between computational problems with output are defined as for decision problems. With this in mind, we get

Lemma 2.25. *Let* (Γ, R) *and* (Σ, S) *be two finite presentations with* $\langle \Gamma \mid R \rangle \cong \langle \Sigma \mid S \rangle$. *Then* $\mathsf{WSP}(\Gamma, R)$ *is logspace many-one reducible to* $\mathsf{WSP}(\Sigma, S)$.

Proof. Let f be an isomorphism from $\langle \Gamma \mid R \rangle$ to $\langle \Sigma \mid S \rangle$. There exists a homomorphism $g : F(\Gamma) \to F(\Sigma)$ that induces f, and similarly, there exists a homomorphism $h : F(\Sigma) \to F(\Gamma)$ that induces f^{-1}. Assume we have an algorithm for solving $\mathsf{WSP}(\Sigma, S)$ and we want to use this algorithm for solving $\mathsf{WSP}(\Gamma, R)$. So, let $w = a_1 \cdots a_n \in (\Gamma \cup \Gamma^{-1})^*$. We have $w = 1$ in $\langle \Gamma \mid R \rangle$ if and only if $g(w) = 1$ in $\langle \Sigma \mid S \rangle$. Hence, if our algorithm for $\mathsf{WSP}(\Sigma, R)$ outputs "no" on input $g(w)$, then we can output "no" on input w. Otherwise, the algorithm for $\mathsf{WSP}(\Sigma, S)$ outputs a representation

$$g(w) = \prod_{i=1}^{\ell} c_i r_i c_i^{-1}$$

in $F(\Sigma)$, where $r_i \in S \cup S^{-1}$. Hence, in $F(\Gamma)$, we have

$$h(g(w)) = h(g(a_1)) \cdots h(g(a_n)) = \prod_{i=1}^{\ell} h(c_i) h(r_i) h(c_i)^{-1}. \tag{2.2}$$

Note that in the group $\langle \Gamma \mid R \rangle$ we have $h(r) = 1$ for all $r \in S$. Hence, in $F(\Gamma)$ we can write

$$h(r_i) = \prod_{j=1}^{k_i} d_{i,j} \, p_{i,j} \, d_{i,j}^{-1}$$

and thus

$$h(c_i) h(r_i) h(c_i)^{-1} = \prod_{j=1}^{k_i} h(c_i) d_{i,j} \, p_{i,j} \, d_{i,j}^{-1} h(c_i)^{-1}$$

with $p_{i,j} \in R \cup R^{-1}$. Here, the words $d_{i,j}$ and the numbers k_i do not depend on the input w. By plugging these identities into (2.2) and renaming the resulting word, we obtain a representation

$$h(g(w)) = h(g(a_1)) \cdots h(g(a_n)) = \prod_{i=1}^{k} d_i \, p_i \, d^{-1}. \qquad (2.3)$$

In the group $\langle \Gamma \mid R \rangle$ we also have $h(g(a)) = a$ for all $a \in \Gamma \cup \Gamma^{-1}$. Hence, we can write in $F(\Gamma)$

$$h(g(a_i))^{-1} a_i = \prod_{j=1}^{\ell_i} e_{i,j} q_{i,j} e_{i,j}^{-1} =: w_i$$

with $q_{i,j} \in R \cup R^{-1}$. Again, the words $e_{i,j}$ and the numbers ℓ_i do not depend on the input w. One can easily check that

$$h(g(a_n))^{-1} \cdots h(g(a_1))^{-1} a_1 \cdots a_n$$
$$= w_n (a_n^{-1} w_{n-1} a_n)(a_n^{-1} a_{n-1}^{-1} w_{n-2} a_{n-1} a_n) \cdots (a_n^{-1} \cdots a_2^{-1} w_1 a_2 \cdots a_n)$$

in $F(\Gamma)$. Finally, we get

$$w = a_1 \cdots a_n = \prod_{i=1}^{k} d_i \, p_i \, d^{-1} w_n (a_n^{-1} w_{n-1} a_n) \cdots (a_n^{-1} \cdots a_2^{-1} w_1 a_2 \cdots a_n),$$

which is the desired representation for w. The above process can be clearly accomplished by a logspace transducer.   □

By Lemma 2.25, it is justified to speak of the word search problem for the group G, briefly WSP(G).

Further results on the word search problem can be found in [131]. Examples of groups, for which the word search problem can be solved in polynomial time, are automatic groups and finitely generated nilpotent groups. For automatic groups, the standard proof that every automatic group is finitely presented and has a quadratic Dehn function [57] yields a polynomial time algorithm for the word search problem. For finitely generated nilpotent groups it has been shown in [62] that a finitely generated nilpotent group of class c has a Dehn function of growth $\mathcal{O}(n^{c+1})$. This result is shown in [62] by reducing a word w with $w = 1$ to the empty word by a sequence of rewrite steps with a system of the kind used in the proof of Proposition 2.22. This sequence of rewrite steps is explicitly constructed in [62] and the construction can be carried out in polynomial time. Finally, the constructed sequence of rewrite steps directly translates into a van Kampen diagram for w.

Chapter 3
Algorithms on Compressed Words

In this chapter, we will study circuits over a structure whose domain is the set Γ^* of all finite words over a finite alphabet. The operations used in the circuit include the binary concatenation and all symbols from Γ as constants. Such circuits are also called *straight-line programs* (SLPs). A straight-line program can be also viewed as a context-free grammar [82] that produces only a single word. Note that, as remarked in Sect. 1.3.7, the term "straight-line program" is also used for arithmetic circuits. In this book, we use the term "straight-line program" exclusively for circuits over free monoids. We will also consider extensions of straight-line programs, where some additional operations on words may be used in circuits (the so-called CSLPs and PCSLPs).

A main topic in this chapter will be algorithms that receive as input a word, which is not given explicitly but as a straight-line program. The main result of this chapter states that for two given straight-line programs one can check in polynomial time whether they produce the same words. For further results on SLPs see [59, 114, 142, 150].

3.1 Straight-Line Programs and Extensions

Let us start with the definition of a PCSLP, which is the most general word circuits that we consider in this book. Here "SLP" abbreviates "straight-line program," and "PC" stands for "projection and cut."

Definition 3.1 (projection). For alphabets $\Delta \subseteq \Gamma$ let us define the projection homomorphism $\pi_\Delta : \Gamma^* \to \Delta^*$ by $\pi_\Delta(a) = \varepsilon$ for $a \in \Gamma \setminus \Delta$ and $\pi_\Delta(a) = a$ for $a \in \Delta$. Strictly speaking, we should denote this homomorphism by $\pi_{\Gamma,\Delta}$, but the alphabet Γ will be always clear from the context.

Definition 3.2 (cut-operator). For natural numbers $i, j \in \mathbb{N}$ and an arbitrary alphabet Γ, we define the mapping $c_{i,j} : \Gamma^* \to \Gamma^*$ by $c_{i,j}(w) = w[i : j]$ for all $w \in \Gamma^*$.

M. Lohrey, *The Compressed Word Problem for Groups*, SpringerBriefs in Mathematics, 43
DOI 10.1007/978-1-4939-0748-9_3, © Markus Lohrey 2014

Definition 3.3 (PCSLP, CSLP, SLP, C-expression, PC-expression). Let Γ be a finite alphabet. We define the following structures with domain Γ^*:

$$\mathscr{A}_\Gamma = (\Gamma^*, \circ, \varepsilon, (a)_{a \in \Gamma}, (\pi_\Delta)_{\Delta \subseteq \Gamma}, (c_{i,j})_{i,j \in \mathbb{N}})$$

$$\mathscr{B}_\Gamma = (\Gamma^*, \circ, \varepsilon, (a)_{a \in \Gamma}, (c_{i,j})_{i,j \in \mathbb{N}})$$

$$\mathscr{C}_\Gamma = (\Gamma^*, \circ, \varepsilon, (a)_{a \in \Gamma})$$

Here, $\circ$ denotes the binary concatenation operator on words. A *PCSLP* over the alphabet Γ is a $\mathscr{A}_\Gamma$-circuit. A *CSLP* over the alphabet Γ is a $\mathscr{B}_\Gamma$-circuit. An *SLP* over the alphabet Γ is a $\mathscr{C}_\Gamma$-circuit. A *C-expression* is $\mathscr{B}_\Gamma$-expression. A *PC-expression* is an $\mathscr{A}_\Gamma$-expression.

PCSLPs and its restrictions will be denoted with letters $\mathbb{A}, \mathbb{B}, \mathbb{C}, \cdots$. For a (PC)SLP $\mathbb{A} = (V, S, \mathrm{rhs})$ over the alphabet Γ we will also write $(V, \Gamma, S, \mathrm{rhs})$ to make the alphabet explicit. Occasionally, we will consider PCSLPs without a start variable. For such a circuit $\mathbb{A}$, $\mathsf{val}(\mathbb{A})$ is not defined.

PCSLPs are the only circuits in this book, which are based on a structure with infinitely many operations (due to the cut-operators $c_{i,j}$). Recall that when we defined the size of a circuit we had to fix a size $\mathsf{size}(f)$ for each function symbol. In case the underlying structure contains only finitely many operations (which is the case for $\mathscr{C}_\Gamma$), we can take $\mathsf{size}(f) = 1$ for all f. For the structure $\mathscr{A}_\Gamma$ let us define $\mathsf{size}(c_{i,j}) = \lceil \log_2(i) \rceil + \lceil \log_2(j) \rceil$, which is the number of bits needed to specify the numbers i and j.

We will always write $e[i : j]$ instead of $c_{i,j}(e)$ in the right-hand sides of PCSLPs. We will use the abbreviations $X[: i]$ for $X[1 : i]$ and $X[i :]$ for $X[i : l]$, where X is a variable with $l = |\mathsf{val}_\mathbb{A}(X)|$, in the right-hand sides. One may also allow exponential expressions of the form A^i for $A \in V$ and $i \in \mathbb{N}$ in the right-hand sides of PCSLPs. Such an expression can be replaced by $\lceil \log_2(i) \rceil$ many ordinary right-hand sides. For instance, the definition $A := a^{10}$ can be replaced by $A_0 := a$, $A_1 := A_0^2$, $A_2 := A_1^2$, $A_3 := A_2^2$, $A := A_3 A_1$. Hence, if we represent exponents in binary notation, then exponential expressions can be eliminated by a logspace transducer.

Example 3.4. We consider the PCSLP $\mathbb{A} = (\{A, B, C, D, E\}, \{a, b, c\}, E, \mathrm{rhs})$ with rhs defined as follows:

$$\mathrm{rhs}(A) = ab \qquad\qquad \mathrm{rhs}(B) = ac \qquad\qquad \mathrm{rhs}(C) = BA$$

$$\mathrm{rhs}(D) = \pi_{\{a,c\}}(C)\pi_{\{b,c\}}(C) \qquad \mathrm{rhs}(E) = D[2 : 4]$$

Then we have

$$\mathsf{val}(A) = ab \qquad \mathsf{val}(B) = ac \qquad\qquad \mathsf{val}(C) = acab$$

$$\mathsf{val}(D) = acacb \qquad \mathsf{val}(E) = \mathsf{val}(\mathbb{A}) = cac$$

The size of the PCSLP is the sum of the sizes of all right-hand sides:

$$|\text{rhs}(A)| = 2 \qquad |\text{rhs}(B)| = 2 \qquad\qquad |\text{rhs}(C)| = 2$$

$$|\text{rhs}(D)| = 4 \qquad |\text{rhs}(E)| = 1 + 1 + 2 = 4$$

and therefore $|\mathbb{A}| = 14$.

CSLPs were called *composition systems* in [60][1], *collage systems* in [98], and *interval grammars* in [71]. Here, we prefer the name CSLP in order to be more consistent in our terminology. Readers familiar with context-free grammars will notice that an SLP $\mathbb{A}$ can be seen as a context-free grammar that generates the single word $\text{val}(\mathbb{A})$. If $\text{rhs}(A) = \alpha$, we also write (as usual for context-free grammars) $A \to \alpha$. Some early references on SLPs are [59, 139, 142].

Definition 3.5 (derivation tree). The *derivation tree* of the SLP $\mathbb{A} = (V, \Gamma, S, \text{rhs})$ is a finite rooted ordered tree, where every node is labeled with a symbol from $V \cup \Gamma$. The root is labeled with the start variable S and every node that is labeled with a symbol from Γ is a leaf of the derivation tree. A node that is labeled with a variable A with $\text{rhs}(A) = \alpha_1 \cdots \alpha_n$ (where $\alpha_1, \ldots, \alpha_n \in V \cup \Gamma$) has n children that are labeled from left to right with $\alpha_1, \ldots, \alpha_n$.

Recall the definition of a circuit in normal form (Definition 1.36), which also applies to PCSLPs, CSLPs, and SLPs. In particular, we allow a right-hand side to be ε. For SLPs that evaluate to a nonempty word, we can forbid the empty word as a right-hand side. An SLP in normal form without the empty word on a right-hand side is a context-free grammar in Chomsky normal form.

Definition 3.6 (Chomsky normal form). An SLP $\mathbb{A} = (V, \Gamma, S, \text{rhs})$ is in *Chomsky normal form*, if for all $A \in V$, $\text{rhs}(A)$ belongs to Γ or has the form BC with $B, C \in V$.

Example 3.7. Consider the SLP $\mathbb{A}$ over the terminal alphabet $\{a, b\}$ with variables $A_1, \ldots, A_7$, where $\text{rhs}(A_1) = a$, $\text{rhs}(A_2) = b$, and $\text{rhs}(A_i) = A_{i-1} A_{i-2}$ for $3 \le i \le 7$. The start variable is A_7. Then $\text{val}(\mathbb{A}) = abaababaabaab$. The SLP $\mathbb{A}$ is in Chomsky normal form and $|\mathbb{A}| = 12$.

We will make heavy use of the Chomsky normal form in this book. Therefore, let us outline an efficient algorithm for transforming a given SLP into Chomsky normal form.

Proposition 3.8. *For a given SLP* $\mathbb{A} = (V, \Gamma, S, \text{rhs})$ *with* $\text{val}(\mathbb{A}) \ne \varepsilon$ *one can compute in time* $\mathcal{O}(|\mathbb{A}|)$ *(on a RAM) an SLP* $\mathbb{B}$ *in Chomsky normal form such that* $\text{val}(\mathbb{A}) = \text{val}(\mathbb{B})$.

---

[1] The formalism in [60] differs in some minor details from CSLPs.

Proof. The proof is similar to the transformation of a context-free grammar into Chomsky normal form and is divided into 4 steps:

Step 1. First, we compute in linear time an enumeration $A_1, A_2, \ldots, A_n$ of V such that if A_i is smaller than A_j with respect to the hierarchical order, then $i < j$, i.e., we compute a topological sorting of V; see, e.g., [42].

Step 2. Using the topological sorting, we can compute the set of variables $E = \{A \in V \mid \mathsf{val}_\mathbb{A}(A) = \varepsilon\}$ in linear time: We set a bit $e(A)$ to zero for each $A \in V$. Then we go over the sequence $A_1, \ldots, A_n$. If $\mathrm{rhs}(A_i)$ only consists of variables and for each variable A_j in $\mathrm{rhs}(A_i)$ we have $e(A_j) = 1$ (this includes the case that $\mathrm{rhs}(A_i) = \varepsilon$), then we set $e(A_i) := 1$. At the end, we have $E = \{A \in V \mid e(A) = 1\}$. Then, we set $V := V \setminus E$ and replace every occurrence of a variable from E in a right-hand side by the empty word. The resulting SLP has no empty right-hand sides.

Step 3. Next, we replace every occurrence of a terminal symbol $a \in \Gamma$ in a right-hand side $\mathrm{rhs}(A) \neq a$ by a new variable A_a with $\mathrm{rhs}(A_a) = a$. At this point, every right-hand side is either a single terminal symbol or a nonempty word over the set of variables V.

Step 4. Right-hand sides of length at least 3 are split into right-hand sides of length 2 by introducing additional variables. For this, we have to go only once over all right-hand sides; hence linear time suffices.

Step 5. Finally, we remove right-hand sides consisting of a single variable (the so-called chain productions). Let $C := \{A \in V \mid \mathrm{rhs}(A) \in V\}$. Then, we go over the set C in topological order and thereby redefine $\mathrm{rhs}(A) := \mathrm{rhs}(\mathrm{rhs}(A))$ for every $A \in C$. The resulting SLP is in Chomsky normal form. $\square$

A simple induction shows that for every SLP $\mathbb{A}$ of size m one has $|\mathsf{val}(\mathbb{A})| \leq \mathcal{O}(3^{m/3})$ [31, proof of Lemma 1]. On the other hand, it is straightforward to define an SLP $\mathbb{B}$ in Chomsky normal form of size $2n$ such that $|\mathsf{val}(\mathbb{B})| \geq 2^n$. Hence, an SLP can be seen as a compressed representation of the word it generates, and exponential compression rates can be achieved in this way.

3.2   Algorithms for Words Represented by SLPs

In this section, we will see some simple polynomial time algorithms for words that are represented by SLPs. In the next section, we will show that CSLPs and under certain restrictions also PCSLPs can be transformed in polynomial time into equivalent SLPs. Hence, all algorithmic problems from this section can be solved in polynomial time for CSLPs and PCSLPs (under some restrictions) as well.

Proposition 3.9. *The following time bounds hold for the RAM model:*

(1) Given an SLP $\mathbb{A} = (V, \Sigma, S, \mathrm{rhs})$, one can compute all alphabets $\mathsf{alph}(\mathsf{val}_\mathbb{A}(A))$ for $A \in V$ in time $\mathcal{O}(|\mathbb{A}|)$.

(2) *Given an SLP* $\mathbb{A} = (V, \Sigma, S, \text{rhs})$, *one can compute all lengths* $|\text{val}_\mathbb{A}(A)|$ *for* $A \in V$ *in time* $\mathcal{O}(|\mathbb{A}|)$.

(3) *Given an SLP* $\mathbb{A}$ *and a number* $1 \le i \le |\text{val}(\mathbb{A})|$, *one compute the symbol* $\text{val}(\mathbb{A})[i]$ *in time* $\mathcal{O}(|\mathbb{A}|)$ *(this problem is in fact* **P**-*complete [107]).*

(4) *Given an SLP* $\mathbb{A}$ *over the terminal alphabet* Γ *and a free monoid homomorphism* $\rho : \Gamma^* \to \Sigma^*$, *one can compute an SLP* $\mathbb{B}$ *such that* $\text{val}(\mathbb{B}) = \rho(\text{val}(\mathbb{A}))$ *in time* $\mathcal{O}(|\mathbb{A}| \cdot \max\{|\rho(a)| \mid a \in \Gamma\})$. *Moreover, this computation is also possible by a logspace transducer.*

(5) *Given an SLP* $\mathbb{A}$ *over the terminal alphabet* Σ *and a symbol* $a \in \Sigma$, *one can compute the number* $\max(\{0\} \cup \{p \mid \text{val}(\mathbb{A})[p] = a\})$ *(i.e., the maximal position on which* $\text{val}(\mathbb{A})$ *contains an* a) *in time* $\mathcal{O}(|\mathbb{A}|)$.

(6) *Given an SLP* $\mathbb{A}$ *over the terminal alphabet* Σ, *a subalphabet* $\Gamma \subseteq \Sigma$, *and a number* $1 \le i \le |\pi_\Gamma(\text{val}(\mathbb{A}))|$, *one can compute the minimal position* $1 \le p \le |\text{val}(\mathbb{A})|$ *with* $|\pi_\Gamma(\text{val}(\mathbb{A})[1 : p])| = i$ *(i.e., the position of the* i-*th occurrence of a letter from* Γ *in* $\text{val}(\mathbb{A})$) *in time* $\mathcal{O}(|\mathbb{A}|)$.

Proof. By Proposition 3.8, we can assume that $\mathbb{A} = (V, \Sigma, S, \text{rhs})$ is in Chomsky normal form. First, we compute in linear time an enumeration $A_1, A_2, \ldots, A_n$ of V such that if A_i is smaller than A_j with respect to the hierarchical order, then $i < j$; see also the proof of Proposition 3.8.

For (1), let us assume that $\Sigma = \{a_1, \ldots, a_m\}$. A subset of Σ can be represented by a bit vector of length m, and the union of two subsets can be computed as the bitwise OR of the corresponding bit vectors. Note that the bitwise OR is an elementary operation in our RAM model; see Sect. 1.3.8. Hence, we store for each nonterminal $A_i \in V$ a bit vector b_i of length m, which is initially set to the 0-vector. Then we go over all nonterminals in the order $A_1, A_2, \ldots, A_n$. If $\text{rhs}(A_i) = a_j$, then we set the j-th bit of b_i to 1. If $\text{rhs}(A_i) = A_j A_k$ with $j, k < i$, then we set b_i to the bitwise OR of b_j and b_k.

Basically, the same algorithm works for (2), except that b_i now stores a natural number. If $\text{rhs}(A_i) = a_j$, then we set $b_i := 1$. If $\text{rhs}(A_i) = A_j A_k$ with $j, k < i$, then we set $b_i := b_j + b_k$. Note that the number of bits of each number b_i is bounded by $\mathcal{O}(\log(|\text{val}_\mathbb{A}(A)|)) \le \mathcal{O}(|\mathbb{A}|)$.

For (3), the algorithm walks down in the derivation tree for $\mathbb{A}$ to position i. By (2), we can assume that for every $A \in V$, the length $|\text{val}_\mathbb{A}(A)|$ is available. At each stage, the algorithm stores a number p and a variable A of $\mathbb{A}$ such that $1 \le p \le |\text{val}_\mathbb{A}(A)|$. Initially, $p = i$, and A is the start variable of $\mathbb{A}$. If $\text{rhs}(A) = BC$, then there are two cases: If $1 \le p \le |\text{val}_\mathbb{A}(B)|$, then we continue with position p and the variable B. Otherwise, we have $|\text{val}_\mathbb{A}(B)| + 1 \le p \le |\text{val}_\mathbb{A}(A)|$, and we continue with position $p - |\text{val}_\mathbb{A}(B)|$ (this number can be computed using the algorithm from the previous point) and variable C. Finally, if $\text{rhs}(A) = a$, then we must have $p = 1$ and we output the terminal symbol a.

For (4) we only have to replace in each right-hand side of $\mathbb{A}$ every occurrence of a terminal symbol by $\rho(a)$.

For (5) let $p_{a,i} = \max(\{0\} \cup \{k \mid \text{val}_\mathbb{A}(A_i)[k] = a\})$ for each $1 \le i \le n$. We compute all values $p_{a,i}$ for $1 \le i \le n$. First, we compute for each variable A_i the alphabet $\mathsf{alph}(\mathsf{val}_\mathbb{A}(A_i))$ (represented by a bit vector according to the proof of (1)) and the length $|\mathsf{val}_\mathbb{A}(A_i)|$ (see (2)) in time $\mathcal{O}(|\mathbb{A}|)$. Then, we compute the positions $p_{a,i}$ as follows: If $\mathsf{rhs}(A_i) = b \in \Sigma$, we set $p_{a,i} = 1$ if $b = a$ and $p_{a,i} = 0$ otherwise. For a variable A_i with $\mathsf{rhs}(A_i) = A_j A_k$ $(j, k < i)$ we set

$$
p_{a,i} = \begin{cases} 0 & \text{if } a \notin \mathsf{alph}(\mathsf{val}_\mathbb{A}(A_i)) \\ p_{a,j} & \text{if } a \in \mathsf{alph}(\mathsf{val}_\mathbb{A}(A_i)) \setminus \mathsf{alph}(\mathsf{val}_\mathbb{A}(A_k)) \\ |\mathsf{val}_\mathbb{A}(A_j)| + p_{a,k} & \text{if } a \in \mathsf{alph}(\mathsf{val}_\mathbb{A}(A_k)). \end{cases}
$$

Finally, the proof of (6) is similar to the proof of (3). By (2) and (4) we can assume that for every $A \in V$ the lengths $|\mathsf{val}_\mathbb{A}(A)|$ and $|\pi_\Gamma(\mathsf{val}_\mathbb{A}(A))|$ are available. We start with setting a variable p to 0 and another variable q to i (where i is from (6)). Then we walk down in the derivation tree for $\mathbb{A}$. If we are currently at a variable $A \in V$ and $\mathsf{rhs}(A) = BC$, then there are two cases: If $1 \leq q \leq |\pi_\Gamma(\mathsf{val}_\mathbb{A}(B))|$, then we continue at variable B and do not change the values of p and q. On the other hand, if $|\pi_\Gamma(\mathsf{val}_\mathbb{A}(B))| < q \leq |\pi_\Gamma(\mathsf{val}_\mathbb{A}(A))|$, then we set $p := p + |\mathsf{val}_\mathbb{A}(B)|$ and $q := q - |\pi_\Gamma(\mathsf{val}_\mathbb{A}(B))|$ and continue at variable C. Finally, if we arrive at a variable A with $\mathsf{rhs}(A) \in \Sigma$, then we terminate and output the current value of p.

This concludes the proof of the proposition. $\qquad\square$

Point (4) from Proposition 3.9 can be extended to mappings computed by deterministic rational transducers, as long as we only require the output SLP to be computed in polynomial time (and not logspace). A *deterministic rational transducer* $\mathscr{T}$ is a tuple $\mathscr{T} = (Q, \Sigma, \Gamma, \delta, q_0, F)$, where

- Q is a finite set of states,
- Σ is the (finite) input alphabet,
- Γ is the (finite) output alphabet,
- $\delta \subseteq Q \times \Sigma \times \Gamma^* \times Q$ is a finite set of transitions such that for every $q \in Q$ and $a \in \Sigma$ there exists at most one transition of the form (q, a, w, p) in δ,
- $q_0 \in Q$ is the initial state, and
- $F \subseteq Q$ is the set of final states.

We define $\hat{\delta}$ as the smallest subset of $Q \times \Sigma^* \times \Gamma^* \times Q$ such that:

- $(q, \varepsilon, \varepsilon, q) \in \hat{\delta}$ for all $q \in Q$,
- $\delta \subseteq \hat{\delta}$, and
- If $(q, u, v, p), (p, u', v', r) \in \hat{\delta}$, then also $(q, uu', vv', r) \in \hat{\delta}$.

Then, for every state $p \in Q$ and every input word $u \in \Sigma^*$, there is at most one state $q \in Q$ and at most one word $v \in \Gamma^*$ such that $(p, u, v, q) \in \hat{\delta}$. Hence, we can define a partial mapping $[\![\mathscr{T}]\!] : \Sigma^* \to \Gamma^*$ by (i) $[\![\mathscr{T}]\!](u) = v$ if there is a final state $q \in F$ such that $(q_0, u, v, q) \in \hat{\delta}$ and (ii) $[\![\mathscr{T}]\!](u) =$ undefined in all other cases. We define the *size* of the transducer $\mathscr{T} = (Q, \Sigma, \Gamma, \delta, q_0, F)$ as $|\mathscr{T}| = |Q| + \sum_{(q,a,w,p) \in \delta} (|w| + 1)$.

Theorem 3.10. *From a given SLP $\mathbb{A}$ over the terminal alphabet Σ and a given deterministic rational transducer $\mathcal{T}$ with input alphabet Σ, one can compute in time $\mathcal{O}(|\mathbb{A}| \cdot |\mathcal{T}|)$ an SLP $\mathbb{B}$ such that* $\mathsf{val}(\mathbb{B}) = [\![\mathcal{T}]\!](\mathsf{val}(\mathbb{A}))$.

Proof. Let $\mathbb{A} = (V, \Sigma, S, \mathrm{rhs})$, which is without loss of generality in Chomsky normal form, and $\mathcal{T} = (Q, \Sigma, \Gamma, \delta, q_0, F)$. In a first step we compute inductively over the hierarchical order of $\mathbb{A}$ for every variable $A \in V$ and every state $p \in Q$ the unique state $s(p, A) \in Q$ such that $(p, \mathsf{val}(A), v, s(p, A)) \in \hat{\delta}$ for some word $v \in \Gamma^*$. If such a word v does not exist, we set $s(p, A) = $ undefined. If $\mathrm{rhs}(A) = a \in \Sigma$, then there is at most one transition $(p, a, w, q) \in \delta$ and we set $s(p, A) = q$ if such a transition exists. If $\mathrm{rhs}(A) = BC$ and for every state $q \in Q$ the states $s(q, B)$ and $s(q, C)$ are already defined, then we set $s(p, A) = s(s(p, B), C)$ for all $p \in Q$ (if $s(p, B)$ is undefined, or $q = s(p, B)$ is defined but $s(q, C)$ is undefined, then clearly $s(p, A)$ is undefined too). This first phase needs time $\mathcal{O}(|V| \cdot |Q|)$.

Now we can define the SLP $\mathbb{B}$. The variables of $\mathbb{B}$ are all pairs $(p, A) \in Q \times V$ such that $s(p, A)$ is defined. The right-hand sides for these variables are defined as follows: Let $A \in V$, $p \in Q$, and $s(p, A) = q$. If $\mathrm{rhs}(A) = a \in \Sigma$, then there is exactly one transition $(p, a, w, q) \in \delta$ and we set $\mathrm{rhs}((p, A)) = w$. If $\mathrm{rhs}(A) = BC$, then $s(p, B) = r$ must be defined. We set $\mathrm{rhs}((p, A)) = (p, B)(r, C)$. This second phase needs time $\mathcal{O}(|V| \cdot \sum_{(q,a,w,p) \in \delta}(|w| + 1))$. $\square$

We will also need the following simple related result.

Theorem 3.11. *For a given SLP $\mathbb{A}$ and a given nondeterministic finite automaton $\mathscr{A} = (Q, \Sigma, \delta, q_0, F)$, we can check in time $\mathcal{O}(|\mathbb{A}| \cdot |Q|^3)$ whether* $\mathsf{val}(\mathbb{A}) \in L(\mathscr{A})$.

Proof. We can assume that $Q = \{1, \ldots, n\}$ and $q_0 = 1$. This allows to represent the set of transitions of the automaton $\mathscr{A}$ by Boolean matrices M_a, one for each symbol $a \in \Sigma$. The matrix M_a has a 1 at entry (i, j) if and only if $(i, a, j) \in \delta$. We can now evaluate the SLP $\mathbb{A} = (V, \Sigma, S, \mathrm{rhs})$ (which is without loss of generality in Chomsky normal form) over Boolean matrices. Formally, for every variable $A \in V$ of $\mathbb{A}$ we define a Boolean matrix M_A as follows: If $\mathrm{rhs}(A) = a \in \Sigma$, then $M_A = M_a$. If $\mathrm{rhs}(A) = BC$, then $M_A = M_B M_C$. Since a multiplication of two Boolean $(n \times n)$-matrices can be done in time n^3, we need time $\mathcal{O}(|V| \cdot n^3)$ to compute all matrices M_A.

Now, $\mathsf{val}(\mathbb{A}) \in L(\mathscr{A})$ if and only if there is a final state $i \in F$ such that entry $(1, i)$ of the matrix M_S is 1. $\square$

In [15, 124] it was shown that there exists a fixed regular language L such that it is **P**-complete to check for a given SLP $\mathbb{A}$ whether $\mathsf{val}(\mathbb{A}) \in L$; see also the remark after Theorem 4.10.

The following lemma will be crucial for our applications of compressed word problems. It generalizes point (4) from Proposition 3.9.

Lemma 3.12. *For a given sequence $\varphi_1, \ldots, \varphi_n$ of free monoid homomorphisms $\varphi_i : \Gamma^* \to \Gamma^*$ $(1 \le i \le n)$ and an SLP $\mathbb{A}$ over the alphabet Γ, one can compute in logarithmic space an SLP $\mathbb{B}$ such that* $\mathsf{val}(\mathbb{B}) = \varphi_1(\varphi_2(\cdots \varphi_n(\mathsf{val}(\mathbb{A})) \cdots))$.

Proof. We compute in logarithmic space SLPs for the words $\varphi_1(\varphi_2(\cdots\varphi_n(a)))$ for all $a \in \Gamma$. Then, one can take the union of all these SLPs (assuming that their variable sets are pairwise disjoint) and replace in the SLP $\mathbb{A}$ every occurrence of a symbol a by the start variable of the SLP for $\varphi_1(\varphi_2(\cdots\varphi_n(a)\cdots))$.

Let us take variables $A_{i,a}$, where $0 \leq i \leq n$ and $a \in \Gamma$, and define

$$\text{rhs}(A_{i,a}) = \begin{cases} a & \text{if } i = 0 \\ A_{i-1,a_1} \cdots A_{i-1,a_m} & \text{if } 0 < i \leq n \text{ and } \varphi_i(a) = a_1 \cdots a_m \end{cases}$$

By induction on i one can easily show that $\text{val}(\mathbb{A}_{i,a}) = \varphi_1(\varphi_2(\cdots\varphi_i(a)\cdots))$. $\square$

We conclude this chapter with a simple problem for words, which becomes difficult for SLPs.

Theorem 3.13. *The following problem is* **NP**-*complete:*

input: Two SLPs $\mathbb{A}$ and $\mathbb{B}$ over the terminal alphabet $\{a, b\}$.
output: Is there a position i such that $\text{val}(\mathbb{A})[i] = \text{val}(\mathbb{B})[i] = b$?

Proof. It is clear that the problem belongs to **NP**: We can guess a position i and then verify $\text{val}(\mathbb{A})[i] = \text{val}(\mathbb{B})[i]$ in polynomial time.

For **NP**-hardness, we use a construction from [112]. We reduce from SUBSET-SUM; see Example 1.42. Let $\overline{w} = (w_1, \ldots, w_n)$ be a tuple of natural numbers and let $t \in \mathbb{N}$; all numbers are encoded in binary. For a bit vector $\overline{x} \in \{0, 1\}^n$ of length n let us define $\overline{x} \cdot \overline{w} = x_1 w_1 + x_2 w_2 + \cdots + x_n w_n$. Note that SUBSETSUM is the question, whether there exists a bit vector $\overline{x} \in \{0, 1\}^n$ with $\overline{x} \cdot \overline{w} = t$. Let $\overline{1}_k$ be the constant-1 vector $(1, 1, \ldots, 1)$ of length k, let $\overline{w}_k = (w_1, \ldots, w_k)$, and let $s_k = \overline{1}_k \cdot \overline{w}_k = w_1 + \cdots + w_k$ for $1 \leq k \leq n$. Finally, let $s = s_n = w_1 + w_2 + \cdots + w_n$ and define the word

$$S(\overline{w}) = \prod_{\overline{x} \in \{0,1\}^n} a^{\overline{x} \cdot \overline{w}} b a^{s - \overline{x} \cdot \overline{w}}.$$

Here, the product $\prod_{\overline{x} \in \{0,1\}^n}$ means that we concatenate all words $a^{\overline{x} \cdot \overline{w}} b a^{s - \overline{x} \cdot \overline{w}}$ for $\overline{x} \in \{0, 1\}^n$ and the order of concatenation is the lexicographic order on $\{0, 1\}^n$, where the right-most bit has the highest significance. For example, we have

$$S(2, 3, 5) = ba^{10} \, a^2 ba^8 \, a^3 ba^7 \, a^5 ba^5 \, a^5 ba^5 \, a^7 ba^3 \, a^8 ba^2 \, a^{10} b$$

$$= ba^{12} ba^{11} ba^{12} ba^{10} ba^{12} ba^{11} ba^{12} b.$$

Let us show that there is an SLP $\mathbb{A}$ of size polynomial in $\sum_{i=1}^n \log(w_i)$ for the word $S(\overline{w})$. Note that $\sum_{i=1}^n \log(w_i)$ is roughly the length of the binary encoding of the tuple $\overline{w}$. The SLP $\mathbb{A}$ contains variables $A_1, \ldots, A_n$ with

$$\text{rhs}(A_1) = ba^{s+w_1} b$$

$$\text{rhs}(A_{k+1}) = A_k a^{s-s_k+w_{k+1}} A_k \quad (1 \leq k \leq n-1).$$

Here we use binary coded numbers as exponents. These powers can be easily removed by ordinary right-hand sides as remarked earlier. To see that $\mathbb{A}$ can be computed by a logspace transducer from $w_1, \ldots, w_n$, it suffices to compute all numbers $s_1, \ldots, s_n$ in logspace. This is an instance of the iterated addition problem (compute the sum of a given tuple of binary encoded numbers), which belongs to the circuit complexity class $\mathbf{TC}^0$. Every function in $\mathbf{TC}^0$ can be computed by a logspace transducer; see [162] for details.

Let A_n be the start variable of $\mathbb{A}$. We prove by induction on k that

$$\mathsf{val}_{\mathbb{A}}(A_k) = \left(\prod_{\overline{x} \in \{0,1\}^k \setminus \{\overline{1}_k\}} a^{\overline{x} \cdot \overline{w}_k} b a^{s - \overline{x} \cdot \overline{w}_k} \right) a^{s_k} b.$$

The case $k = 1$ is clear, since $\mathsf{val}_{\mathbb{A}}(A_1) = b a^{s + w_1} b = b a^s a^{s_1} b = a^0 b a^{s-0} a^{s_1} b$. For $k + 1 \leq n$ we obtain the following:

$$\left(\prod_{\overline{x} \in \{0,1\}^{k+1} \setminus \{\overline{1}_{k+1}\}} a^{\overline{x} \cdot \overline{w}_{k+1}} b a^{s - \overline{x} \cdot \overline{w}_{k+1}} \right) a^{s_{k+1}} b$$

$$= \underbrace{\left(\prod_{\overline{x} \in \{0,1\}^k} a^{\overline{x} \cdot \overline{w}_k} b a^{s - \overline{x} \cdot \overline{w}_k} \right)}_{\mathsf{val}_{\mathbb{A}}(A_k) a^{s - s_k}} \underbrace{\left(\prod_{\overline{x} \in \{0,1\}^k \setminus \{\overline{1}_k\}} a^{\overline{x} \cdot \overline{w}_k + w_{k+1}} b a^{s - \overline{x} \cdot \overline{w}_k - w_{k+1}} \right) a^{w_{k+1}} a^{s_k} b}_{a^{w_{k+1}} \mathsf{val}_{\mathbb{A}}(A_k)}$$

$$= \mathsf{val}_{\mathbb{A}}(A_k) a^{s - s_k + w_{k+1}} \mathsf{val}_{\mathbb{A}}(A_k) = \mathsf{val}_{\mathbb{A}}(A_{k+1}).$$

For $k = n$ we finally get

$$\mathsf{val}(\mathbb{A}) = \mathsf{val}_{\mathbb{A}}(A_n) = \prod_{\overline{x} \in \{0,1\}^n} a^{\overline{x} \cdot \overline{w}} b a^{s - \overline{x} \cdot \overline{w}} = S(\overline{w}).$$

Our second SLP $\mathbb{B}$ satisfies

$$\mathsf{val}(\mathbb{B}) = (a^t b a^{s-t})^{2^n}.$$

It is straightforward to construct such an SLP in logspace from the binary representations of s and t. Clearly, there exists a bit vector $\overline{x} \in \{0,1\}^n$ with $\overline{x} \cdot \overline{w} = t$ if and only if there is a position i with $\mathsf{val}(\mathbb{A})[i] = \mathsf{val}(\mathbb{B})[i] = b$. This concludes the proof. $\square$

3.3 Transforming CSLPs and PCSLPs into SLPs

CSLPs and PCSLPs are sometimes more convenient than SLPs. On the other hand, in this section we will see that CSLPs can be transformed in polynomial time into SLPs. The same holds for PCSLPs if we restrict applications of projection operators suitably.

The following result was shown by Hagenah in his PhD thesis [71] (in German); see also [151].

Theorem 3.14. *From a given CSLP* $\mathbb{A} = (V_{\mathbb{A}}, \Gamma, \text{rhs}_{\mathbb{A}})$ *(without start variable) in normal form with n variables, one can compute in time* $\mathcal{O}(n^2)$ *an SLP* $\mathbb{B} = (V_{\mathbb{B}}, \Gamma, \text{rhs}_{\mathbb{B}})$ *(without start variable) of size* $\mathcal{O}(n^2)$ *such that* $V_{\mathbb{A}} \subseteq V_{\mathbb{B}}$ *and* $\text{val}_{\mathbb{B}}(X) = \text{val}_{\mathbb{A}}(X)$ *for all* $X \in V_{\mathbb{A}}$.

Proof. Let us write V and rhs for $V_{\mathbb{A}}$ and $\text{rhs}_{\mathbb{A}}$, respectively, in the following. Let h be the height of $\mathbb{A}$. The idea is to push cut-operators downwards (towards smaller variables in the hierarchical order). First we compute the lengths of all words $\text{val}_{\mathbb{A}}(A)$ for $A \in V$. This is possible in time $\mathcal{O}(n)$ as for SLPs; see Proposition 3.9. Choose a variable $A \in V$ such that $\text{rhs}(A) = B[i : j]$, but every smaller variable C (smaller with respect to the hierarchical order) has no cut-operator in its right-hand side. We show that we can eliminate the cut-operator in $\text{rhs}(A)$ and thereby obtain a CSLP $\mathbb{A}' = (V', \Gamma, S, \text{rhs}')$ in normal form such that $V \subseteq V'$, $|V' \setminus V| \leq 2h$, $\text{val}_{\mathbb{A}}(A) = \text{val}_{\mathbb{A}'}(A)$ for all $A \in V$, and the height of $\mathbb{A}'$ is bounded by the height of $\mathbb{A}$. By iterating this transformation, we can eliminate all cut-operators and transform $\mathbb{A}$ into an equivalent SLP of size $|\mathbb{A}| + \mathcal{O}(h|V|)$ (the elimination of each cut-operator increases the size by $\mathcal{O}(h)$).

If $\text{rhs}(B) = a$ for $a \in \Gamma$, then we set $\text{rhs}(A) = a$ if $i = j = 1$ and $\text{rhs}(A) = \varepsilon$ otherwise. Now assume that $\text{rhs}(B) = CD$ (recall that we assume that $\mathbb{A}$ is in normal form). Let $m = |\text{val}(C)|$ and $n = |\text{val}(D)|$. There are three cases:

Case 1. $j \leq m$. We redefine $\text{rhs}(A) = C[i, j]$.
Case 2. $i \geq n + 1$. We redefine $\text{rhs}(A) = D[i - m, j - m]$.
Case 3. $i \leq m$ and $j > m$. In this case we introduce two new variables C' and D', redefine $\text{rhs}(A) = C'D'$, and set $\text{rhs}(C') = C[i :]$ and $\text{rhs}(D') = D[: i - m]$.

By iterating this process, we arrive at one of the following two situations:

 (i) After several applications of Case 1 and 2 we have $\text{rhs}(A) = X[k, l]$ and $\text{rhs}(X) = a \in \Gamma$ for some variable $X \in V$.
 (ii) We arrive at Case 3 for the first time.

In situation (i) we are done (and have not introduced any new variables). So assume that we arrive in situation (ii). We have introduced two new variables C' and D'. Let us deal with C' (with D' we deal analogously). We have set $\text{rhs}(C') = C[i :]$. If $\text{rhs}(C) = a$ for $a \in \Gamma$, then we set $\text{rhs}(C') = a$ if $i = 1$ and $\text{rhs}(C) = \varepsilon$ otherwise. Now assume that $\text{rhs}(C) = EF$ for $E, F \in V$. Let $m = |\text{val}(E)|$ and $n = |\text{val}(F)|$. We distinguish two cases:

Case 1. $i \leq m$. We introduce a new variable E', redefine $\mathrm{rhs}(C') = E'F$, set $\mathrm{rhs}(E') = E[i\,:]$, and continue with E'.

Case 2. $i > m$. We redefine $\mathrm{rhs}(C') = F[:i-m]$ and continue with C'.

By iterating this process we finally eliminate the cut-operator. Note that in each step at most one new variable is introduced (in Case 1). Therefore, at most d variables are introduced. Since we have to do an analogous procedure for D', we introduce at most $2h$ new variables in total. Clearly, our process does not increase the height of the CSLP. Moreover, the resulting CSLP is again in normal form except for variables X with $\mathrm{rhs}(X) = \varepsilon$. But these variables can be eliminated at the end. This proves the theorem. $\square$

Theorem 3.14 cannot be generalized to PCSLPs, simply because the size of a smallest SLP that is equivalent to a given PCSLP may be exponential in the size of the PCSLP. Here is an example.

Example 3.15. For $n \geq 1$ let us define the PCSLP

$$\mathbb{A}_n = (\{A_1, \ldots, A_n, A_{n+1}\}, \Gamma_n, A_1, \mathrm{rhs})$$

with $\Gamma_n = \{a_1, \ldots, a_n\}$, $\mathrm{rhs}(A_i) = \pi_{\Delta_i}(A_{i+1})A_{i+1}$ for $1 \leq i \leq n$, and $\mathrm{rhs}(A_{n+1}) = a_1 a_2 \cdots a_n$, where $\Delta_i = \{a_1, \ldots, a_n\} \setminus \{a_i\}$. This PCSLP generates the word

$$\mathsf{val}(\mathbb{A}_n) = \prod_{\Delta \subseteq \Gamma_n} \pi_\Delta(a_1 a_2 \cdots a_n), \tag{3.1}$$

where all subsets $\Delta \subseteq \Gamma_n$ are listed in a kind of lexicographic order (the precise order is not important for the following argument). Let $\mathbb{B}_n$ be an SLP with $\mathsf{val}(\mathbb{A}_n) = \mathsf{val}(\mathbb{B}_n)$. By [31, Lemma 3], $\mathsf{val}(\mathbb{B}_n)$ contains at most $k \cdot |\mathbb{B}_n|$ many different factors of length k. But from (3.1) we see that $\mathsf{val}(\mathbb{A}_n) = \mathsf{val}(\mathbb{B}_n)$ contains at least $\binom{n}{\lceil n/2 \rceil}$ many factors of length $\lceil n/2 \rceil$. Hence, $|\mathbb{B}_n|$ must be at least $\lceil n/2 \rceil^{-1} \cdot \binom{n}{\lceil n/2 \rceil}$, which is exponential in n.

Nevertheless, under certain restrictions we can transform a given PCSLP into an equivalent SLP.

Lemma 3.16. *Let p be a constant. Then there exists a polynomial time algorithm for the following problem:*

input: Finite alphabets $\Gamma_1, \ldots, \Gamma_p$ and a PCSLP $\mathbb{A} = (V_\mathbb{A}, \Gamma, \mathrm{rhs}_\mathbb{A})$ (without start variable) over $\Gamma = \bigcup_{i=1}^p \Gamma_i$ such that $\Delta \in \{\Gamma_1, \ldots, \Gamma_p\}$ for every subexpression of the form $\pi_\Delta(\alpha)$ that appears in a right-hand side of $\mathbb{A}$.
output: An SLP $\mathbb{B} = (V_\mathbb{B}, \Gamma, \mathrm{rhs}_\mathbb{B})$ (without start variable) over Γ such that $V_\mathbb{A} \subseteq V_\mathbb{B}$ and $\mathsf{val}_\mathbb{A}(X) = \mathsf{val}_\mathbb{B}(X)$ for every $X \in V_\mathbb{A}$.

Proof. Since for a CSLP an equivalent SLP can be constructed in polynomial time by Theorem 3.14, it suffices to construct in polynomial time a CSLP $\mathbb{B}$ with the desired properties from the statement of the lemma. Let

$$\mathscr{C} = \{\bigcap_{i \in K} \Gamma_i \mid K \subseteq \{1, \ldots, p\}\} \cup \{\Gamma\}.$$

Note that $\mathscr{C}$ has constant size. Let $V_{\mathbb{B}} = \{X_\Delta \mid X \in V_{\mathbb{A}}, \Delta \in \mathscr{C}\}$ be the set of variables of $\mathbb{B}$. We identify $X \in V_{\mathbb{A}}$ with $X_\Gamma \in V_{\mathbb{B}}$. The right-hand side mapping $\mathrm{rhs}_{\mathbb{B}}$ will be defined in such a way that $\mathrm{val}_{\mathbb{B}}(X_\Delta) = \pi_\Delta(\mathrm{val}_{\mathbb{A}}(X))$.

If $\mathrm{rhs}_{\mathbb{A}}(X) = a \in \Gamma$, then we set $\mathrm{rhs}_{\mathbb{B}}(X_\Delta) = \pi_\Delta(a) \in \{\varepsilon, a\}$. If $\mathrm{rhs}_{\mathbb{A}}(X) = YZ$, then we set $\mathrm{rhs}_{\mathbb{B}}(X_\Delta) = Y_\Delta Z_\Delta$. If $\mathrm{rhs}_{\mathbb{A}}(X) = \pi_\Theta(Y)$ with $\Theta \in \{\Gamma_1, \ldots, \Gamma_p\}$, then we set $\mathrm{rhs}_{\mathbb{B}}(X_\Delta) = Y_{\Delta \cap \Theta}$. Note that $\Delta \cap \Theta \in \mathscr{C}$.

Finally, consider the case $\mathrm{rhs}_{\mathbb{A}}(X) = Y[i : j]$. We set $\mathrm{rhs}_{\mathbb{B}}(X_\Delta) = Y_\Delta[k : \ell]$, where $k = |\pi_\Delta(\mathrm{val}(Y)[: i - 1])| + 1$ and $\ell = |\pi_\Delta(\mathrm{val}(Y)[: j])|$. These lengths can be computed in polynomial time as follows: Implicitly, we have already computed a CSLP, which generates the word $\mathrm{val}(Y)$. Hence, by adding a single definition, we obtain a CSLP for the word $\mathrm{val}(Y)[: i - 1]$. Using Theorem 3.14 we can transform this CSLP in polynomial time into an equivalent SLP, which can be transformed in polynomial time into an SLP for the word $\pi_\Delta(\mathrm{val}(Y)[: i-1])$ by Proposition 3.9(4). From this SLP we can compute the length $|\pi_\Delta(\mathrm{val}(Y)[: i - 1])|$ in polynomial time by Proposition 3.9 (the SLP for the word $\mathrm{val}(Y)[: i - 1]$ is not used in the further computation). The length $|\pi_\Delta(\mathrm{val}(Y)[: j])|$ can be computed similarly. Since the size of $\mathscr{C}$ is constant, the above construction works in polynomial time. $\square$

In the proof of Lemma 3.16, it is crucial that p is a fixed constant, i.e., not part of the input. Otherwise the construction would lead to an exponential blowup. Example 3.15 shows that this is unavoidable.

By Theorem 3.14, all algorithmic problems for SLP-represented words considered in Sect. 3.2 can be solved in polynomial time for CSLP-represented words as well. The same applies to PCSLPs under the restriction for the alphabet from Lemma 3.16.

3.4   Compressed Equality Checking

The most basic task for SLP-compressed words is equality checking: Given two SLPs $\mathbb{A}$ and $\mathbb{B}$, does $\mathrm{val}(\mathbb{A}) = \mathrm{val}(\mathbb{B})$ hold? Clearly, a simple decompress-and-compare strategy is very inefficient. It takes exponential time to compute $\mathrm{val}(\mathbb{A})$ and $\mathrm{val}(\mathbb{B})$. Nevertheless a polynomial time algorithm exists. This was independently discovered by Hirshfeld, Jerrum, and Moller [80]; Mehlhorn, Sundar, and Uhrig [128]; and Plandowski [139].

Theorem 3.17 ([80, 128, 139]). *The following problem belongs to* **P**:

input: Two SLPs $\mathbb{A}$ *and* $\mathbb{B}$.
question: Does $\mathsf{val}(\mathbb{A}) = \mathsf{val}(\mathbb{B})$ *hold?*

A natural generalization of checking equality of two words is pattern matching. In the classical *pattern matching problem* it is asked for given words p (usually called the pattern) and t (usually called the text) whether p is a factor of t. There are several linear time algorithms for this problem on uncompressed words, most notably the well-known Knuth-Morris-Pratt algorithm [99]. It is therefore natural to ask whether a polynomial time algorithm for pattern matching on SLP-compressed words exists; this problem is sometimes called *fully compressed pattern matching* and is defined as follows:

input: Two SLPs $\mathbb{A}$ and $\mathbb{B}$.
question: Is $\mathsf{val}(\mathbb{A})$ a factor of $\mathsf{val}(\mathbb{B})$?

The first polynomial time algorithm for fully compressed pattern matching was presented in [96]; further improvements with respect to the running time were achieved in [59, 89, 106, 130]. In this book, we will only need the weaker Theorem 3.17.

In the rest of this section, we will prove Theorem 3.17. Since the result is so fundamental for the rest of the book, we will also analyze the precise running time of our algorithm (using the RAM model from Sect. 1.3.8). Our algorithm will use Jeż's recompression technique [89]. This technique was used in [89] to give the currently fastest algorithm for fully compressed pattern matching. Basically, our algorithm can be seen as a simplification of the algorithm from [89].[2] We start with several definitions.

Let $s \in \Sigma^+$ be a nonempty word over a finite alphabet Σ. We define the word $\mathsf{block}(s)$ as follows: Assume that $s = a_1^{n_1} a_2^{n_2} \cdots a_k^{n_k}$ with $a_1, \ldots, a_k \in \Sigma$, $a_i \neq a_{i+1}$ for all $1 \leq i < k$, and $n_i > 0$ for all $1 \leq i \leq k$. Then $\mathsf{block}(s) = a_1^{(n_1)} a_2^{(n_2)} \cdots a_k^{(n_k)}$, where $a_1^{(n_1)}, a_2^{(n_2)}, \ldots, a_k^{(n_k)}$ are new symbols. For instance, for $s = aabbbaccb$ we have $\mathsf{block}(s) = a^{(2)} b^{(3)} a^{(1)} c^{(2)} b^{(1)}$. For the symbol $a^{(1)}$ we will simply write a. Let us set $\mathsf{block}(\varepsilon) = \varepsilon$.

For a partition $\Sigma = \Sigma_l \uplus \Sigma_r$ we denote with $s[\Sigma_l, \Sigma_r]$ the word that is obtained from s by replacing every occurrence of a factor ab in s with $a \in \Sigma_l$ and $b \in \Sigma_r$ by the new symbol $\langle ab \rangle$. For instance, for $s = abcbabcad$ and $\Sigma_l = \{a, c\}$ and $\Sigma_r = \{b, d\}$ we have $s[\Sigma_l, \Sigma_r] = \langle ab \rangle \langle cb \rangle \langle ab \rangle c \langle ad \rangle$. Since two different occurrences of factors from $\Sigma_l \Sigma_r$ must occupy disjoint sets of positions in s, the word $s[\Sigma_l, \Sigma_r]$ is well defined.[3]

---

[2]Using a refined analysis, Jeż moreover achieves a better running time in [89] compared to our algorithm.

[3]More formally, one can define the semi-Thue system $R = \{(ab, \langle ab \rangle) \mid a \in \Sigma_l, b \in \Sigma_r\}$. It is Noetherian and confluent, and we have $s[\Sigma_l, \Sigma_r] = \mathsf{NF}_R(s)$.

Obviously, for all words $s, t \in \Sigma^*$ we have

$$(s = t \iff \mathsf{block}(s) = \mathsf{block}(t)) \quad \text{and} \quad (s = t \iff s[\Sigma_l, \Sigma_r] = t[\Sigma_l, \Sigma_r]).$$

$$(3.2)$$

In the rest of this section, we assume that all SLPs $\mathbb{A} = (V, \Sigma, S, \mathrm{rhs})$ are in a kind of generalized Chomsky normal form: We require that for every variable $A \in V$, $\mathrm{rhs}(A)$ is either of the form $u \in \Sigma^+$, uBv with $u, v \in \Sigma^*$ and $B \in V$ or $uBvCw$ with $u, v, w \in \Sigma^*$ and $B, C \in V$. In other words, every right-hand side is nonempty and contains at most two occurrences of variables. In particular, we only consider SLPs that produce nonempty word. This is not a crucial restriction for checking the equality $\mathsf{val}(\mathbb{A}) = \mathsf{val}(\mathbb{B})$, since we can first check easily in polynomial time whether $\mathsf{val}(\mathbb{A})$ or $\mathsf{val}(\mathbb{B})$ produce the empty string.

Following Jeż's recompression technique [89], our strategy for checking an equality $\mathsf{val}(\mathbb{A}) = \mathsf{val}(\mathbb{B})$ is to compute from $\mathbb{A}$ and $\mathbb{B}$ two SLPs $\mathbb{A}'$ and $\mathbb{B}'$ such that $\mathsf{val}(\mathbb{A}') = (\mathsf{block}(\mathsf{val}(\mathbb{A})))[\Sigma_l, \Sigma_r]$ and $\mathsf{val}(\mathbb{B}') = (\mathsf{block}(\mathsf{val}(\mathbb{B})))[\Sigma_l, \Sigma_r]$,

---

Algorithm 1: CompressBlocks($\mathbb{A}$)

---

input : SLP $\mathbb{A} = (V, \Sigma, S, \mathrm{rhs})$
output: SLP $\mathbb{B}$ with $\mathsf{val}(\mathbb{B}) = \mathsf{block}(\mathsf{val}(\mathbb{A}))$
let $A_1, A_2, \dots, A_m$ be an enumeration of V in hierarchical order
for $i := 1$ **to** m **do**
  if $\mathrm{rhs}(A_i) = u \in \Sigma^+$ **then**
    | $\mathrm{rhs}(A_i) := \mathsf{block}(u)$
  else if $\mathrm{rhs}(A_i) = uBv$ *with* $u, v \in \Sigma^*$ *and* $B \in V$ **then**
    | $\mathrm{rhs}(A_i) := \mathsf{block}(u)\, B\, \mathsf{block}(v)$
  else
    let $\mathrm{rhs}(A_i) = uBvCw$ with $u, v, w \in \Sigma^*$ and $B, C \in V$
    $\mathrm{rhs}(A_i) := \mathsf{block}(u)\, B\, \mathsf{block}(v)\, C\, \mathsf{block}(w)$
  end
end
for $i := 1$ **to** $m - 1$ **do**
  if $\mathrm{rhs}(A_i)$ *is of the form* $a^{(k)}$ **then**
    replace every occurrence of A_i in a right-hand side by $a^{(k)}$
    remove A_i from the SLP
  else
    let $\mathrm{rhs}(A_i)$ be of the form $a^{(k)} \alpha b^{(l)}$
    if $\alpha = \varepsilon$ **then**
      replace every occurrence of A_i in a right-hand side by $a^{(k)} b^{(l)}$
      remove A_i from the SLP
    else
      replace every occurrence of A_i in a right-hand side by $a^{(k)} A_i b^{(l)}$
      $\mathrm{rhs}(A_i) := \alpha$
    end
  end
  replace every factor $c^{(j)} c^{(k)}$ ($c \in \{a, b\}$) in a right-hand side by $c^{(j+k)}$
end

---

where the partition is chosen in such a way that the total length $|\text{val}(\mathbb{A}')| + |\text{val}(\mathbb{B}')|$ is bounded by $c \cdot (|\text{val}(\mathbb{A})| + |\text{val}(\mathbb{B})|)$ for some constant $c < 1$. This process is iterated. After at most $\log(|\text{val}(\mathbb{A})| + |\text{val}(\mathbb{B})|) \in \mathcal{O}(|\mathbb{A}| + |\mathbb{B}|)$ many iterations, it must terminate with two SLPs, one of which produces a string of length one. Checking equality of the two words produced by these SLPs is easy. The main difficulty of this approach is to bound the size of the two SLPs during this process.

Lemma 3.18. *Given an SLP* $\mathbb{A}$, $\text{alph}(\text{block}(\text{val}(\mathbb{A})))$ *contains at most* $|\mathbb{A}|$ *many different symbols.*

Proof. A block in $s = \text{val}(\mathbb{A})$ is a maximal factor of the form $s[i, j] = a^n$ for $a \in \Sigma$ (maximal in the sense that either $i = 1$ or $s[i - 1] \neq a$ and either $j = |s|$ or $s[j + 1] \neq a$). We have to show that s contains at most $|\mathbb{A}|$ different blocks. To every block a^n that is a factor of s we can assign a unique variable A such that a^n is a factor of $\text{val}_\mathbb{A}(A)$, but a^n is not a factor of any word $\text{val}_\mathbb{A}(B)$, where B is a variable that occurs in $\text{rhs}(A)$. But there are at most $|\text{rhs}(A)|$ such blocks. Summing over all variables yields the lemma. $\qquad\square$

Lemma 3.19. *Given an SLP* $\mathbb{A}$, *one can compute in time* $\mathcal{O}(|\mathbb{A}|)$ *an SLP* $\mathbb{B}$ *such that* $\text{val}(\mathbb{B}) = \text{block}(\text{val}(\mathbb{A}))$. *Moreover, if* $\mathbb{A}$ *has m variables, then* $\mathbb{B}$ *has at most m variables and* $|\mathbb{B}| \leq |\mathbb{A}| + 4m$.

Proof. Let $\mathbb{A} = (V, \Sigma, S, \text{rhs})$ (where every right-hand side is nonempty and contains at most two variables) and let $A_1, A_2, \ldots, A_m$ be an enumeration of V that respects the hierarchical order of $\mathbb{A}$: If A_i occurs in $\text{rhs}(A_j)$, then $i < j$. Such an enumeration can be produced in time $\mathcal{O}(|\mathbb{A}|)$; it is equivalent to the problem of computing a topological sorting of an acyclic graph; see, e.g., [42, Sect. 22.4]. We can assume that $A_m = S$. Let a_i (respectively, b_i) be the first (respectively, last) symbol of $\text{val}_\mathbb{A}(A_i)$. Moreover, let φ be the homomorphism with $\varphi(a^{(k)}) = a^k$ for all $a \in \Sigma$ and $k \geq 1$.

Consider Algorithm 1 (CompressBlock). It gradually modifies the input SLP $\mathbb{A}$. Let $\mathbb{A}_0$ be the SLP before the second for-loop is executed for the first time, and let $\mathbb{A}_i$ be the SLP produced after $1 \leq i \leq m-1$ iterations of the second for-loop. Finally let $\mathbb{B} = \mathbb{A}_{m-1}$ be the output SLP. Hence, $\varphi(\text{val}(\mathbb{A}_0)) = \varphi(\text{block}(\text{val}(\mathbb{A}))) = \text{val}(\mathbb{A})$. By induction on $0 \leq i \leq m - 1$ we prove the following invariants:

(i) For all $1 \leq j \leq i$ we have that if $\text{block}(\text{val}_\mathbb{A}(A_j))$ has length at most two, then A_j does no longer occur in $\mathbb{A}_i$. On the other hand, if $\text{block}(\text{val}_\mathbb{A}(A_j)) = a^{(k)}ub^{(l)}$ with $u \neq \varepsilon$, then $\text{val}_{\mathbb{A}_i}(A_j) = u$.

(ii) For all $i + 1 \leq j \leq m$, we have $\varphi(\text{val}_{\mathbb{A}_i}(A_j)) = \text{val}_\mathbb{A}(A_j)$ and $\text{rhs}_{\mathbb{A}_i}(A_j)$ does not contain a factor of the form $a^{(p)}a^{(q)}$. Moreover, every occurrence of a variable A_k with $1 \leq k \leq i$ in $\text{rhs}_{\mathbb{A}_i}(A_j)$ has a symbol of the form $a_k^{(p)}$ on its left and a symbol of the form $b_k^{(q)}$ on its right.

For $i = 0$, (i) and (ii) clearly hold. Now assume that (i) and (ii) hold for some $0 \leq i \leq m - 2$ and consider the $(i + 1)$-th iteration of the for-loop. Only the right-hand sides of variables from $A_{i+1}, \ldots, A_m$ are modified. Hence, (i) still holds for all variables $A_1, \ldots, A_i$ after the $(i + 1)$-th iteration. In order to prove the remaining points, note that all variables in $\mathrm{rhs}_{\mathbb{A}_i}(A_{i+1})$ are among $\{A_1, \ldots, A_i\}$ and hence satisfy point (i) after the i-th iteration. Moreover, A_{i+1} satisfies (ii) after the i-th iteration. We get $\mathsf{val}_{\mathbb{A}_i}(A_{i+1}) = \mathsf{block}(\mathsf{val}_{\mathbb{A}}(A_{i+1}))$ and $\mathrm{rhs}_{\mathbb{A}_i}(A_{i+1})$ either has the form $a^{(p)} = \mathsf{block}(\mathsf{val}_{\mathbb{A}}(A_{i+1}))$ or $a^{(k)} \alpha b^{(l)}$, where $a^{(k)}$ (respectively, $b^{(l)}$) is the first (respectively, last) symbol of $\mathsf{block}(\mathsf{val}_{\mathbb{A}}(A_{i+1}))$. Using this, it is straightforward to verify (i) for A_{i+1} as well as (ii) for all variables $A_{i+2}, \ldots, A_m$.

Note that $\mathbb{B}$ has at most m variables (no new variables are introduced) and that the length of a right-hand side of $\mathbb{A}$ increases by at most 4 in the construction of $\mathbb{B}$: Every right-hand side of $\mathbb{A}$ contains at most two variables, and for each occurrence of a variable in a right-hand side we add one symbol on the left and one symbol on the right. Hence, we get $|\mathbb{B}| \leq |\mathbb{A}| + 4m$.   □

Let us now analyze words $s[\Sigma_l, \Sigma_r]$ for a partition $\Sigma = \Sigma_\ell \uplus \Sigma_r$.

Lemma 3.20. *For a word $s \in \Sigma^*$ which does not contain factor of the form aa with $a \in \Sigma$ (i.e., $\mathsf{block}(s) = s$) there exists a partition $\Sigma = \Sigma_\ell \uplus \Sigma_r$ such that $|s[\Sigma_l, \Sigma_r]| \leq (3|s| + 1)/4$.*

Proof. The following probabilistic argument is given in [90]. Let $s = a_1 a_2 \cdots a_n$. We put each symbol $a \in \Sigma$ with probability $1/2$ into Σ_l and with probability $1/2$ into Σ_r. Hence, for every position $1 \leq i \leq n-1$, the probability that $a_i a_{i+1} \in \Sigma_l \Sigma_r$ is $1/4$ (here, we need $a_i \neq a_{i+1}$). By linearity of expectations (for which we do not need independence), the expected number of positions $1 \leq i \leq n - 1$ with $a_i a_{i+1} \in \Sigma_l \Sigma_r$ is $(n-1)/4$. Hence, there exists a partition $\Sigma = \Sigma_\ell \uplus \Sigma_r$ for which there are at least $(n - 1)/4$ many positions $1 \leq i \leq n - 1$ with $a_i a_{i+1} \in \Sigma_l \Sigma_r$. For this partition, we get $|s[\Sigma_l, \Sigma_r]| \leq n - (n - 1)/4 = (3n + 1)/4$.   □

Next, we show that for an SLP-represented string one can compute a partition as in Lemma 3.20 in linear time.

Lemma 3.21. *Assume that the alphabet Σ is an initial segment of the natural numbers. Given an SLP $\mathbb{A}$ over the terminal alphabet Σ such that $s := \mathsf{val}(\mathbb{A}) = \mathsf{block}(\mathsf{val}(\mathbb{A}))$, one can compute in time $\mathcal{O}(|\mathbb{A}|)$ a partition $\Sigma = \Sigma_l \uplus \Sigma_r$ such that $|s[\Sigma_l, \Sigma_r]| \leq (3|s| + 1)/4$.*

Proof. Let $n = |\mathbb{A}|$. First note that the word $s = \mathsf{val}(\mathbb{A})$ contains at most n many different diagrams (= factors of length 2). The argument is similar to the proof of Lemma 3.18: To each diagram ab of s we can assign a unique variable A such that ab is a factor of $\mathsf{val}_{\mathbb{A}}(A)$ but ab is not a factor of any word $\mathsf{val}_{\mathbb{A}}(B)$, where B is a variable that occurs in $\mathrm{rhs}(A)$. But there are at most $|\mathrm{rhs}(A)| - 1$ such factors. Summing over all variables yields the statement.

Consider Algorithm 2 (CountingDiagrams), which computes for each letter $a \in \Sigma$ a list $R(a)$ that contains all pairs (b, v) such that the diagram ab occurs in s and the number of positions $1 \leq i \leq |s| - 1$ such that $s[i, i + 1] = ab$ is equal to v. By the remark in the preceding paragraph, the total number of entries in all lists $R(a)$ ($a \in \Sigma$) is bounded by n. The algorithm first computes a topological sorting

$A_1, \ldots, A_m$ of the variables with respect to the hierarchical order. Then, it computes for every $A_i \in V$ the number v_i of occurrences of A_i in the derivation tree of $\mathbb{A}$. This part of the algorithm is taken from [65]. Next, an array B of at most m entries is computed. This array contains all triples (a, b, i), where $\text{rhs}(A_i) = A_j A_k$ (for some $j, k < i$), the last symbol of $\text{val}_{\mathbb{A}}(A_j)$ is a, and the first symbol of $\text{val}_{\mathbb{A}}(A_k)$ is b. Note that for each occurrence of ab in s there exists a unique node v in the derivation tree of $\mathbb{A}$ such that the occurrence of ab is below v, but it is neither below the left child of v nor below the right child of v. If v is labeled with A_i and

---

Algorithm 2: CountingDigramms

Data: SLP $\mathbb{A} = (V, \Sigma, S, \text{rhs})$ with $\text{val}(\mathbb{A}) = \text{block}(\text{val}(\mathbb{A}))$
transform $\mathbb{A}$ into Chomsky normal form in time $\mathcal{O}(|\mathbb{A}|)$
let $A_1, A_2, \ldots, A_m$ be an enumeration of V in hierarchical order
for $i := 1$ *to* $m - 1$ **do**
| $v_i := 0$
end
$v_m := 1$
for $i := m$ ***downto*** 2 **do**
| **if** $\text{rhs}(A_i) = A_j A_k$ **then**
| | $v_j := v_j + v_i$
| | $v_k := v_k + v_i$
| **end**
end
$c := 0$
for $i := 1$ *to* m **do**
| **if** $\text{rhs}(A_i) = a \in \Sigma$ **then**
| | $\text{first}(i) := a; \text{last}(i) := a$
| **else**
| | let $\text{rhs}(A_i) = A_j A_k$ with $j, k < i$
| | $\text{first}(i) := \text{first}(j); \text{last}(i) := \text{last}(k)$
| | $a := \text{last}(j); b := \text{first}(k)$
| | $c := c + 1$
| | $B[c] := (a, b, i)$
| **end**
end
sort array B using radix sort on the lexicographic order on tuples from $\Sigma \times \Sigma \times \{1, \ldots, m\}$, ignoring the last component, and let C be the resulting array (of length $c \leq m$)
for $a \in \Sigma$ **do**
| $R(a) :=$ empty list
end
for $i := 1$ *to* c **do**
| let $C[i] = (a, b, i)$
| **if** $i = 1$ *or* $C[i - 1]$ *is not of the form* (a, b, j) *for some* j **then**
| | append to $R(a)$ the entry (b, v_i)
| **else**
| | let the last entry in $R(a)$ be (b, v)
| | replace this entry by $(b, v + v_i)$
| **end**
end

Algorithm 3: ComputePartition($\mathbb{A}$)

Data: SLP $\mathbb{A} = (V, \Sigma, S, \text{rhs})$ with $\text{val}(\mathbb{A}) = \text{block}(\text{val}(\mathbb{A}))$
compute list $L(a)$ and $R(a)$ for all $a \in \Sigma$ using Algorithm 2
$\Sigma_l := \emptyset; \Sigma_r := \emptyset; \Gamma := \emptyset$
for $a \in \Sigma$ **do**
  | $l_a := 0; r_a := 0$
end
for $a \in \Sigma$ **do**
  | **if** $r_a \geq l_a$ **then**
  | | $\Sigma_l := \Sigma_l \cup \{a\}$
  | | **for** $(b, v) \in L(a)$ **do**
  | | | $l_b := l_b + v$
  | | **end**
  | | **for** $(b, v) \in R(a)$ **do**
  | | | $l_b := l_b + v$
  | | **end**
  | **else**
  | | $\Sigma_r := \Sigma_r \cup \{a\}$
  | | **for** $(b, v) \in L(a)$ **do**
  | | | $r_b := r_b + v$
  | | **end**
  | | **for** $(b, v) \in R(a)$ **do**
  | | | $r_b := r_b + v$
  | | **end**
  | **end**
end
if $\sum_{a \in \Sigma_r} \sum_{(b,v) \in (\Sigma_l \times \mathbb{N}) \cap R(a)} v > \sum_{a \in \Sigma_l} \sum_{(b,v) \in (\Sigma_r \times \mathbb{N}) \cap R(a)} v$ **then**
  | $(\Sigma_l, \Sigma_r) := (\Sigma_r, \Sigma_l)$
end

$\text{rhs}(A_i) = A_j A_k$, then the last (respectively, first) symbol of A_j (respectively, A_k) must be a (respectively, b). Hence, the total number of occurrences of ab in s is the sum of all numbers v_i such that (a, b, i) occurs in the array R.

In order to compute the lists $R(a)$, the array B is sorted using radix sort on the lexicographic order on $\Sigma \times \Sigma$ (the first two components from B-entries), where the second Σ-component has higher priority. Radix sort is an algorithm that allows to sort an array of m many k-ary numbers having d digits in time $\mathcal{O}(dm + dk)$ on a RAM; see [42]. To apply radix sort in our situation, we need the assumption that the alphabet Σ is an initial segment of natural numbers; see also [90]. Then, radix sort needs time $\mathcal{O}(m + |\Sigma|) \leq \mathcal{O}(|\mathbb{A}|)$ to sort the array B. From the sorted array C it is easy to compute the lists $R(a)$ for all $a \in \Sigma$ in a single pass over C.

Completely analogous to the lists $R(a)$, one can also compute in linear time lists $L(a)$ ($a \in \Sigma$), where $L(a)$ contains all pairs (b, v) such that the diagram ba occurs in s and the number of positions $1 \leq i \leq |s| - 1$ such that $s[i, i + 1] = ab$ is equal to v. Both lists $R(a)$ and $L(a)$ are needed in Algorithm 3 (ComputePartition) (see again [90]) for computing a partition $\Sigma = \Sigma_l \uplus \Sigma_r$. The two sets Σ_l and Σ_r are stored as bit maps, i.e., arrays of length $|\Sigma|$, where the a-th array entry is 0 or 1,

---

Algorithm 4: CompressPairs($\mathbb{A}, \Sigma_l, \Sigma_r$)

Data: SLP $\mathbb{A} = (V, \Sigma, S, \text{rhs})$ with $\text{val}(\mathbb{A}) = \text{block}(\text{val}(\mathbb{A}))$ and a partition $\Sigma = \Sigma_l \uplus \Sigma_r$.

let $A_1, A_2, \ldots, A_m$ be an enumeration of V in hierarchical order

for $i := 1$ **to** $m - 1$ **do**

     if $\text{rhs}(A_i)$ *is of the form* $a\alpha$ *for* $a \in \Sigma_r$ **then**

         replace every occurrence of A_i in a right-hand side by $a A_i$

         $\text{rhs}(A_i) := \alpha$

     end

     if $\text{rhs}(A_i)$ *is of the form* βb *for* $b \in \Sigma_l$ **then**

         replace every occurrence of A_i in a right-hand side by $A_i b$

         $\text{rhs}(A_i) := \beta$

     end

     if $\text{rhs}(A_i) = \varepsilon$ **then**

         replace every occurrence of A_i in a right-hand side by ε

         remove A_i from the SLP

     end

end

replace every occurrence of a factor $ab \in \Sigma_l \Sigma_r$ in a right-hand side by $\langle ab \rangle$

---

depending on whether a belongs to Σ_l (respectively, Σ_r). Here, we need again the assumption that $\Sigma = \{1, \ldots, |\Sigma|\}$.

ComputePartition first computes in a greedy way a partition $\Sigma = \Sigma_l \uplus \Sigma_r$ such that the number of positions $1 \leq i \leq |s|-1$ with $s[i, i+1] \in \Sigma_l \Sigma_r \cup \Sigma_r \Sigma_l$ is at least $(|s|-1)/2$. For this the algorithm goes over all symbols from Σ and makes for every $a \in \Sigma$ a greedy choice. If for the current sets Σ_l and Σ_r the number of occurrences of diagrams from $a\Sigma_r \cup \Sigma_r a$ is at least as large as the number of occurrences of diagrams from $a\Sigma_l \cup \Sigma_l a$, then the algorithm puts a into Σ_l; otherwise a is put into Σ_r. To make this choice, the algorithm stores (and correctly updates) the number of occurrences of diagrams from $a\Sigma_r \cup \Sigma_r a$ (resp., $a\Sigma_l \cup \Sigma_l a$) in the variable r_a (resp., l_a). These variables are updated using the lists $L(a)$ and $R(a)$.

Since the number of positions $1 \leq i \leq |s| - 1$ with $s[i, i + 1] \in \Sigma_l \Sigma_r \cup \Sigma_r \Sigma_l$ is at least $(|s|-1)/2$, there exist at least $(|s|-1)/4$ positions i with $s[i, i + 1] \in \Sigma_l \Sigma_r$ or there exist at least $(|s| - 1)/4$ positions i with $s[i, i + 1] \in \Sigma_r \Sigma_l$. The last if-statement in Algorithm 3 selects the partition which yields more diagram occurrences. Note that the sum $\sum_{a \in \Sigma_r} \sum_{(b,v) \in (\Sigma_l \times \mathbb{N}) \cap R(a)} v$ can be computed in linear time since the sets Σ_l and Σ_r are stored as bit maps and the total length of all lists $R(a)$ is at most n. A similar remark applies to the sum $\sum_{a \in \Sigma_l} \sum_{(b,v) \in (\Sigma_r \times \mathbb{N}) \cap R(a)} v$. $\qquad\square$

Since we have to deal with two SLPs that have to be checked for equality, we need a simple adaptation of Lemma 3.21 for two SLPs.

Lemma 3.22. *There is an algorithm* ComputePartition($\mathbb{A}, \mathbb{B}$) *that computes for a given SLPs $\mathbb{A}$ and $\mathbb{B}$ over the terminal alphabet Σ with $s := \text{val}(\mathbb{A}) = \text{block}(\text{val}(\mathbb{A}))$ and $t := \text{val}(\mathbb{B}) = \text{block}(\text{val}(\mathbb{B}))$ in time $\mathscr{O}(|\mathbb{A}| + |\mathbb{B}|)$ a partition such that $|s[\Sigma_l, \Sigma_r]| + |t[\Sigma_l, \Sigma_r]| \leq 3(|s| + |t|)/4 + 5/4$.*

Algorithm 5: CheckEquality

Data: SLPs $\mathbb{A}$ and $\mathbb{B}$
while $|\mathsf{val}(\mathbb{A})| > 1$ *and* $|\mathsf{val}(\mathbb{B})| > 1$ **do**
   | $\mathbb{A} := \text{CompressBlocks}(\mathbb{A})$
   | $\mathbb{B} := \text{CompressBlocks}(\mathbb{B})$
   | let Γ be the union of the terminal alphabets of $\mathbb{A}$ and $\mathbb{B}$
   | $(\Gamma_l, \Gamma_r) := \text{ComputePartition}(\mathbb{A}, \mathbb{B})$
   | $\mathbb{A} := \text{CompressPairs}(\mathbb{A}, \Gamma_l, \Gamma_r)$
   | $\mathbb{B} := \text{CompressPairs}(\mathbb{B}, \Gamma_l, \Gamma_r)$
end
check whether $\mathsf{val}(\mathbb{A}) = \mathsf{val}(\mathbb{B})$

Proof. By taking the disjoint union of all productions of $\mathbb{A}$ and $\mathbb{B}$ we can easily construct an SLP $\mathbb{C}$ of size $\mathscr{O}(|\mathbb{A}| + |\mathbb{B}|)$ such that $\mathsf{val}(\mathbb{C}) = st =: u$. With Lemma 3.21 we obtain a partition $\Sigma = \Sigma_l \cup \Sigma_r$ such that $|u[\Sigma_l, \Sigma_r]| \leq (3|u| + 1)/4 = (3(|s| + |t|) + 1)/4$. On the other hand, we have $|u[\Sigma_l, \Sigma_r]| \geq |s[\Sigma_l, \Sigma_r]| + |t[\Sigma_l, \Sigma_r]| - 1$. Hence, we get $|s[\Sigma_l, \Sigma_r]| + |t[\Sigma_l, \Sigma_r]| \leq 3(|s| + |t|)/4 + 5/4$. □

Lemma 3.23. *Given an SLP $\mathbb{A}$ over the terminal alphabet Σ such that $\mathsf{val}(\mathbb{A}) = \mathsf{block}(\mathsf{val}(\mathbb{A}))$ and a partition $\Sigma = \Sigma_\ell \cup \Sigma_r$ one can compute in time $\mathscr{O}(|\mathbb{A}|)$ an SLP $\mathbb{B}$ such that $\mathsf{val}(\mathbb{B}) = \mathsf{val}(\mathbb{A})[\Sigma_l, \Sigma_r]$. Moreover, if $\mathbb{A}$ has m variables, then $\mathbb{B}$ has at most m variables and $|\mathbb{B}| \leq |\mathbb{A}| + 4m$.*

Proof. Let $\mathbb{A} = (V, \Sigma, S, \text{rhs})$ and let $A_1, A_2, \ldots, A_m$ be an enumeration of V that respects the hierarchical order of $\mathbb{A}$: If A_i occurs in $\text{rhs}(A_j)$, then $i < j$. We can assume that $S = A_m$. Let a_i (respectively, b_i) be the first (respectively, last) symbol of $\mathsf{val}_\mathbb{A}(A_i)$. Consider Algorithm 4 (CompressPairs) and let $\mathbb{B}$ be the output SLP of this algorithm. The proof that $\mathsf{val}(\mathbb{B}) = \mathsf{val}(\mathbb{A})[\Sigma_l, \Sigma_r]$ is similar to the proof of Lemma 3.19.

For a word $w \in \Sigma^*$ define the words $\mathsf{pop}_l(w)$ and $\mathsf{pop}_r(w)$ as follows:

$$\mathsf{pop}_l(\varepsilon) = \varepsilon \qquad\qquad\qquad \mathsf{pop}_r(\varepsilon) = \varepsilon$$

$$\mathsf{pop}_l(au) = au \text{ for } a \in \Sigma_l, u \in \Sigma^* \qquad \mathsf{pop}_r(ua) = ua \text{ for } a \in \Sigma_r, u \in \Sigma^*$$

$$\mathsf{pop}_l(au) = u \text{ for } a \in \Sigma_r, u \in \Sigma^* \qquad \mathsf{pop}_r(ua) = u \text{ for } a \in \Sigma_l, u \in \Sigma^*$$

Moreover, let $\mathsf{pop}(w) = \mathsf{pop}_l(\mathsf{pop}_r(w)) = \mathsf{pop}_r(\mathsf{pop}_l(w))$. Note that for a single symbol a we have $\mathsf{pop}(a) = \varepsilon$.

Let $\mathbb{A}_i$ be the SLP after i iterations of the for-loop ($0 \leq i \leq m - 1$). Hence, $\mathbb{A}_0 = \mathbb{A}$. By induction on $0 \leq i \leq m - 1$ one can easily prove the following invariants:

(i) For all $1 \leq j \leq i$ we have that if $\mathsf{pop}(\mathsf{val}_\mathbb{A}(A_j)) = \varepsilon$, then variable A_i does no longer occur in $\mathbb{A}_i$. Otherwise, $\mathsf{val}_{\mathbb{A}_i}(A_j) = \mathsf{pop}(\mathsf{val}_\mathbb{A}(A_j))$.

(ii) For all $i + 1 \leq j \leq m$ we have $\mathsf{val}_{\mathbb{A}_i}(A_j) = \mathsf{val}_{\mathbb{A}}(A_j)$. Moreover, every occurrence in $\mathsf{rhs}_{\mathbb{A}_i}(A_j)$ of a variable A_k with $1 \leq k \leq i$ and $a_k \in \Sigma_r$ (respectively, $b_k \in \Sigma_l$) is preceded (respectively, followed) by a_k (respectively, b_k).

Hence, we have $\mathsf{val}(\mathbb{A}) = \mathsf{val}(\mathbb{A}_{m-1})$. Moreover, for every factorization $\alpha A_i \beta$ of a right-hand side from $\mathbb{A}_{m-1}$, we have:

- Either α does not end with a symbol from Σ_l or $\mathsf{val}_{\mathbb{A}_{m-1}}(A_i)$ does not start with a symbol from Σ_r.
- Either β does not start with a symbol from Σ_r or $\mathsf{val}_{\mathbb{A}_{m-1}}(A_i)$ does not end with a symbol from Σ_l.

This implies that for the final SLP $\mathbb{B}$, which results from replacing in all right-hand sides of $\mathbb{A}_m$ all occurrences of factors $ab \in \Sigma_l \Sigma_r$ by $\langle ab \rangle$, we have $\mathsf{val}(\mathbb{B}) = \mathsf{val}(\mathbb{A})[\Sigma_l, \Sigma_r]$ for every variable.

Note that $\mathbb{B}$ has at most m variables (no new variables are introduced) and that the length of a right-hand side of $\mathbb{A}$ increases by at most 4 in the construction of $\mathbb{B}$: Every right-hand side of $\mathbb{A}$ contains at most two variables, and for each occurrence of a variable in a right-hand side we introduce at most one symbol on the left and right. Hence, we get $|\mathbb{B}| \leq |\mathbb{A}| + 4m$. □

Proof of Theorem 3.17. Assume that we have two SLPs $\mathbb{A}$ and $\mathbb{B}$ over the same terminal alphabet Σ and let $m := |\mathsf{val}(\mathbb{A})|$ and $n := |\mathsf{val}(\mathbb{B})|$. Moreover, let k (respectively, l) be the number of variables of $\mathbb{A}$ (respectively, $\mathbb{B}$). Algorithm 5 (CheckEquality) checks whether $\mathsf{val}(\mathbb{A}) = \mathsf{val}(\mathbb{B})$. Correctness of the algorithm follows from observation (3.2). It remains to analyze the running time of the algorithm. By Lemma 3.22, the number of iterations of the while-loop is bounded by $\mathcal{O}(\log(n + m)) \leq \mathcal{O}(|\mathbb{A}| + |\mathbb{B}|)$. Let $\mathbb{A}_i$ and $\mathbb{B}_i$ be the SLPs after i iterations of the while-loop. The number of variables of $\mathbb{A}_i$ (respectively, $\mathbb{B}_i$) is at most k (respectively, l). Hence, by Lemmas 3.19 and 3.23, the size of $\mathbb{A}_i$ (respectively, $\mathbb{B}_i$) can be bounded by $|\mathbb{A}| + 4ki \in \mathcal{O}((|\mathbb{A}| + |\mathbb{B}|)^2)$ (respectively, $|\mathbb{B}| + 4li \in \mathcal{O}((|\mathbb{A}| + |\mathbb{B}|)^2)$). Since the i-th iteration takes time $\mathcal{O}(|\mathbb{A}_i| + |\mathbb{B}_i|)$, the total running time is $\mathcal{O}((|\mathbb{A}| + |\mathbb{B}|)^3)$. □

3.5 2-Level PCSLPs

For our algorithms in Chap. 5, it is useful to consider PCSLPs, which are divided into two layers.

Definition 3.24 (2-level PCSLP). A *2-level PCSLP* is a tuple

$$\mathbb{A} = (\mathsf{Up}, \mathsf{Lo}, \Gamma, S, \mathsf{rhs})$$

such that the following holds:

- Up, Lo, and Γ are pairwise disjoint finite alphabets, $S \in$ Up, and rhs : Up $\cup$ Lo $\to$ PC(Up $\cup$ Lo, Γ).
- The tuple (Up, Lo, S, rhs $\upharpoonright_{\text{Up}}$) is a PCSLP over the terminal alphabet Lo.
- The tuple (Lo, Γ, rhs $\upharpoonright_{\text{Lo}}$) is an SLP (without start variable) over the terminal alphabet Γ.

The set Up (respectively, Lo) is called the set of *upper level variables* (*lower level variables*) of $\mathbb{A}$. Moreover, we set $V = $ Up $\cup$ Lo and call it the set of variables of $\mathbb{A}$. The PCSLP (Up, Lo, S, rhs $\upharpoonright_{\text{Up}}$) is called the *upper part of* $\mathbb{A}$, briefly up($\mathbb{A}$), and the SLP (without start variable) (Lo, Γ, rhs $\upharpoonright_{\text{Lo}}$) is the *lower part of* $\mathbb{A}$, briefly lo($\mathbb{A}$). The upper level evaluation mapping uval$_{\mathbb{A}}$: PC(Up, Lo) $\to$ Lo* of $\mathbb{A}$ is defined as uval$_{\mathbb{A}}$ = val$_{\text{up}(\mathbb{A})}$. The evaluation mapping val$_{\mathbb{A}}$ is defined by val$_{\mathbb{A}}(X)$ = val$_{\text{lo}(\mathbb{A})}$(val$_{\text{up}(\mathbb{A})}(X)$) for $X \in$ Up and val$_{\mathbb{A}}(X)$ = val$_{\text{lo}(\mathbb{A})}(X)$ for $X \in$ Lo. Finally, we set val($\mathbb{A}$) = val$_{\mathbb{A}}(S)$. We define the size of $\mathbb{A}$ as $|\mathbb{A}| = \sum_{X \in V} |\text{rhs}(X)|$.

Example 3.25. Let $\mathbb{A} = (\{F, G, H\}, \{A, B, C, D, E\}, \{a, b, c\}, H, \text{rhs})$ be a two-level PCSLP with Up $= \{F, G, H\}$ and Lo $= \{A, B, C, D, E\}$, where the mapping rhs defined as follows:

$$\text{rhs}(A) = a \qquad\qquad \text{rhs}(B) = b \qquad\qquad \text{rhs}(C) = c$$

$$\text{rhs}(D) = AB \qquad\qquad \text{rhs}(E) = AC$$

$$\text{rhs}(F) = EABCDEA$$

$$\text{rhs}(G) = F[2:6]$$

$$\text{rhs}(H) = \pi_{\{A,C,D\}}(G)$$

Then we have

$$\text{up}(\mathbb{A}) = (\{F, G, H\}, \{A, B, C, D, E\}, H, \text{rhs} \upharpoonright_{\text{Up}}) \text{ and}$$

$$\text{lo}(\mathbb{A}) = (\{A, B, C, D, E\}, \{a, b, c\}, \text{rhs} \upharpoonright_{\text{Lo}}).$$

The uval$_{\mathbb{A}}$-values for the upper level variables are

$$\text{uval}_{\mathbb{A}}(F) = EABCDEA$$

$$\text{uval}_{\mathbb{A}}(G) = ABCDE$$

$$\text{uval}_{\mathbb{A}}(H) = ACD$$

The val$_{\mathbb{A}}$-values for all variables of $\mathbb{A}$ are

$$\text{val}_{\mathbb{A}}(A) = a \qquad \text{val}_{\mathbb{A}}(B) = b \qquad \text{val}_{\mathbb{A}}(C) = c$$

$$\text{val}_{\mathbb{A}}(D) = ab \qquad \text{val}_{\mathbb{A}}(E) = ac$$

$$\text{val}_\mathbb{A}(F) = \text{val}_\mathbb{A}(EABCDEA) = acabcabaca$$

$$\text{val}_\mathbb{A}(G) = \text{val}_\mathbb{A}(ABCDE) = abcabac$$

$$\text{val}(\mathbb{A}) = \text{val}_\mathbb{A}(H) = \text{val}_\mathbb{A}(ACD) = acab$$

Note that $\text{val}_\mathbb{A}(G)$ is different from $\text{val}_\mathbb{A}(F)[2:6] = cabca$

Chapter 4
The Compressed Word Problem

In this chapter, we introduce the main topic of this book, namely the compressed word problem for a finitely generated group. This is the variant of the word problem, where the input word is not written down explicitly, but given by an SLP. Since the input word is given in a more succinct way, the compressed word problem for a group G may have a higher computational complexity than the word problem for G, and in Sect. 4.8 we will see such a group. In Sect. 4.1 we show that the complexity of the compressed word problem is preserved when going to a finitely generated subgroup or a finite extension. Section 4.2 demonstrates that an efficient algorithm for the compressed word problem for a group G leads to efficient algorithms for (ordinary) word problems for various groups derived from G (automorphism groups, semidirect products, and other group extensions). The remaining Sects. 4.3–4.8 study the complexity of the compressed word problem in various classes of groups (finite groups, free groups, finitely generated linear groups, finitely generated nilpotent groups, wreath products).

4.1 The Compressed Word Problem and Basic Closure Properties

Let us start with the definition of the compressed word problem.

Definition 4.1. Let G be a finitely generated group and fix a finite generating set Γ for G. The *compressed word problem* for G with respect to Γ, briefly $\mathsf{CWP}(G, \Gamma)$, is the following decision problem:

input: An SLP $\mathbb{A}$ over the terminal alphabet $\Gamma \cup \Gamma^{-1}$.
question: Does $\mathsf{val}(\mathbb{A}) = 1$ hold in G?

In $\mathsf{CWP}(G, \Gamma)$, the input size is of course the size $|\mathbb{A}|$ of the SLP $\mathbb{A}$. As for the (uncompressed) word problem, the complexity of the compressed word problem does not depend on the chosen generating set.

M. Lohrey, *The Compressed Word Problem for Groups*, SpringerBriefs in Mathematics, DOI 10.1007/978-1-4939-0748-9_4, © Markus Lohrey 2014

Lemma 4.2. *Let G be a finitely generated group and let Γ and Σ be two generating sets. Then $\mathsf{CWP}(G, \Gamma) \leq_m^{\log} \mathsf{CWP}(G, \Sigma)$.*

Proof. Let $G \cong \langle \Gamma \mid R \rangle \cong \langle \Sigma \mid S \rangle$. There exists a homomorphism $h : (\Gamma \cup \Gamma^{-1})^* \to (\Sigma \cup \Sigma^{-1})^*$ with $h(a^{-1}) = h(a)^{-1}$ for all $a \in \Gamma \cup \Gamma^{-1}$ that induces an isomorphism from $\langle \Gamma \mid R \rangle$ to $\langle \Sigma \mid S \rangle$. Hence, for a word $w \in (\Gamma \cup \Gamma^{-1})^*$ we have $w = 1$ in $\langle \Gamma \mid R \rangle$ if and only if $h(w) = 1$ in $\langle \Sigma \mid S \rangle$. The lemma follows, since by Proposition 3.9(4) we can compute from a given SLP $\mathbb{A}$ over $\Gamma \cup \Gamma^{-1}$ in logarithmic space an SLP $\mathbb{B}$ over $\Sigma \cup \Sigma^{-1}$ such that $\mathsf{val}(\mathbb{B}) = h(\mathsf{val}(\mathbb{A}))$. This gives a logspace many-one reduction from $\mathsf{CWP}(G, \Gamma)$ to $\mathsf{CWP}(G, \Sigma)$. $\square$

By Lemma 4.2, we can just speak about the compressed word problem for G, briefly $\mathsf{CWP}(G)$.

Before we consider the compressed word problem in specific groups we prove two preservation results. Recall the reducibility relations from Sect. 1.3.3. By the following simple proposition, the complexity of the compressed word problem is preserved when going to a finitely generated subgroup.

Proposition 4.3. *Assume that H is a finitely generated subgroup of the finitely generated group G. Then $\mathsf{CWP}(H) \leq_m^{\log} \mathsf{CWP}(G)$.*

Proof. Choose a generating set Γ for G that contains a generating set Σ for H. Then for a word $w \in (\Sigma \cup \Sigma^{-1})^*$ we have $w = 1$ in H if and only if $w = 1$ in G. $\square$

By the following result from [116], the complexity of the compressed word problem is also preserved when going to a finite extension.

Theorem 4.4. *Assume that K is a finitely generated subgroup of the group G such that the index $[G : K]$ is finite (hence, G is finitely generated too). Then $\mathsf{CWP}(G) \leq_m^P \mathsf{CWP}(K)$.*

Proof. Let Γ be a finite generating set for K and let Σ be a finite generating set for G. Let $h : (\Sigma \cup \Sigma^{-1})^* \to G$ be the canonical morphism. Let $Kg_1, \ldots, Kg_n$ be a list of the cosets of K, where without loss of generality $g_1 = 1$. Let $\mathscr{A}$ be the coset automaton of K. This is an NFA over the alphabet $\Sigma \cup \Sigma^{-1}$ and with state set $\{Kg_1, \ldots, Kg_n\}$. The initial and final state is $K = Kg_1$ and the triple (Kg_i, a, Kg_j) ($a \in \Sigma \cup \Sigma^{-1}$) is a transition of $\mathscr{A}$ if and only if $Kg_i a = Kg_j$. Note that this automaton accepts a word $w \in (\Sigma \cup \Sigma^{-1})^*$ if and only if $h(w) \in K$. Since by Theorem 3.11 it can be checked in polynomial time whether the word generated by a given SLP is accepted by a given NFA (here, we even have a fixed NFA $\mathscr{A}$), we can check in polynomial time whether $h(\mathsf{val}(\mathbb{A})) \in K$ for a given SLP $\mathbb{A}$.

Now let $\mathbb{A} = (V, \Sigma \cup \Sigma^{-1}, S, \mathrm{rhs})$ be an SLP in Chomsky normal form over the alphabet $\Sigma \cup \Sigma^{-1}$. We want to check whether $\mathsf{val}(\mathbb{A}) = 1$ in G. First, we check in polynomial time whether $h(\mathsf{val}(\mathbb{A})) \in K$. If not, we reject immediately (formally, since we have to construct a polynomial time many-one reduction from $\mathsf{CWP}(G)$

to $\text{CWP}(K)$, we should output some fixed SLP over $\Gamma \cup \Gamma^{-1}$ that evaluates to an element of $K \setminus \{1\}$). Otherwise, we will construct in polynomial time an SLP $\mathbb{B}$ over the generating set $\Gamma \cup \Gamma^{-1}$ of K, which computes the same group element as $\mathbb{A}$.

The set of variables of $\mathbb{B}$ is the set of triples

$$W = \{[g_i, A, g_j^{-1}] \mid A \in V, 1 \leq i, j \leq n, \; g_i h(\text{val}_{\mathbb{A}}(A))g_j^{-1} \in K\}.$$

By the above observation, this set can be computed in polynomial time. Now, let us define the right-hand sides for the variables $[g_i, A, g_j^{-1}] \in W$. First, assume that $\text{rhs}(A) = a$, where $a \in \Sigma \cup \Sigma^{-1}$. Hence, $g_i a g_j^{-1} \in K$, and we set $\text{rhs}([g_i, A, g_j^{-1}]) = w$, where $w \in (\Gamma \cup \Gamma^{-1})^*$ is such that $h(w) = g_i a g_j^{-1}$ (we do not have to compute this word w; it is a fixed word that does not depend on the input). Now assume that $\text{rhs}(A) = BC$. In polynomial time, we can determine the unique k such that $g_i h(\text{val}_{\mathbb{A}}(B))$ belongs to the coset $K g_k$. Thus, $g_i h(\text{val}_{\mathbb{A}}(B))g_k^{-1} \in K$, i.e., $[g_i, B, g_k^{-1}] \in W$. We set

$$\text{rhs}([g_i, A, g_j^{-1}]) = [g_i, B, g_k^{-1}][g_k, C, g_j^{-1}].$$

Note that

$$g_i h(\text{val}_{\mathbb{A}}(A))g_j^{-1} = g_i h(\text{val}_{\mathbb{A}}(B))g_k^{-1} g_k h(\text{val}_{\mathbb{A}}(C))g_j^{-1}.$$

Hence, since $g_i h(\text{val}_{\mathbb{A}}(A))g_j^{-1}$ and $g_i h(\text{val}_{\mathbb{A}}(B))g_k^{-1}$ both belong to the subgroup K, we also have $g_k h(\text{val}_{\mathbb{A}}(C))g_j^{-1} \in K$, i.e., $[g_k, C, g_j^{-1}] \in W$. Finally, let $[g_1, S, g_1^{-1}] = [1, S, 1]$ be the start variable of $\mathbb{B}$. Since we assume that $h(\text{val}(\mathbb{A})) = h(\text{val}_{\mathbb{A}}(S)) \in K$, we have $[1, S, 1] \in W$. It is easy to prove that for every variable $[g_i, A, g_j^{-1}] \in W$, $\text{val}_{\mathbb{B}}([g_i, A, g_j^{-1}])$ represents the group element $g_i h(\text{val}_{\mathbb{A}}(A))g_j^{-1}$. Thus, $\text{val}(\mathbb{A}) = 1$ in G if and only if $\text{val}(\mathbb{B}) = 1$ in K, which is an instance of $\text{CWP}(K)$. This proves the theorem. $\qquad\square$

The reducibility relation $\leq_m^P$ in Theorem 4.4 cannot be replaced by the stronger relation $\leq_m^{\log}$ (unless $\mathbf{P} = \mathbf{L}$) because there exists a finite group G with a $\mathbf{P}$-complete compressed word problem; see Theorem 4.10 (take $K = 1$ in Theorem 4.4).

4.2 From the Compressed Word Problem to the Word Problem

It turns out that an efficient algorithm for the compressed word problem for a group G can be used to solve efficiently the (uncompressed) word problem in certain groups derived from G. Hence, the compressed word problem is useful for the solution of the ordinary word problem. In this section, we present three results of this type. All three results are formulated in terms of certain reducibilities; see Sect. 1.3.3.

Definition 4.5 (Aut(G)). For a group G, Aut(G) denotes the *automorphism group* of G, which consists of all automorphisms of G with composition of functions as the group operation.

Recall from Definition 1.27 the definition of $\leq_{bc}^{\log}$.

Theorem 4.6 (cf [151]). *Let G be a finitely generated group and let H be a finitely generated subgroup of* Aut(G). *Then* WP(H) $\leq_{bc}^{\log}$ CWP(G).

Proof. Let Σ be a finite generating set for G, where without loss of generality $a \in \Sigma$ implies $a^{-1} \in \Sigma$. Let H be generated by the finite set $A \subseteq$ Aut(G), where again $\varphi \in A$ implies $\varphi^{-1} \in A$.

For a given input word $\varphi_1 \varphi_2 \cdots \varphi_n$ (with $\varphi_i \in A$ for $1 \leq i \leq n$ we have to check whether the composition of $\varphi_1, \varphi_2, \ldots, \varphi_n$ (in that order) is the identity isomorphism in order to solve the word problem for H. But this is equivalent to $\varphi_n(\varphi_{n-1}(\cdots \varphi_1(a) \cdots)) = a$ in G for all $a \in \Sigma$.

Since Σ is closed under inverses, every $\varphi \in A$ can be viewed as a homomorphism on Σ^*. Hence, by Lemma 3.12 we can compute in logarithmic space an SLP $\mathbb{A}_a$ over the alphabet Σ such that val($\mathbb{A}_a$) $= \varphi_n(\varphi_{n-1}(\cdots \varphi_1(a) \cdots))a^{-1}$. Thus, the composition of $\varphi_1, \varphi_2, \ldots, \varphi_n$ is the identity if and only if for all $a \in \Sigma$, val($\mathbb{A}_a$) $= 1$ in G. Since $|\Sigma|$ is a constant in our consideration, we obtain WP(H) $\leq_{bc}^{\log}$ CWP(G). $\square$

It should be noted that there are finitely generated (even finitely presented) groups G, where Aut(G) is not finitely generated; see, e.g., [105]. Therefore, we restrict in Theorem 4.6 to a finitely generated subgroup of Aut(G).

Definition 4.7 (semidirect product). Let K and Q be groups and let $\varphi : Q \to$ Aut(K) be a group homomorphism. Then the *semidirect product* $K \rtimes_\varphi Q$ is the group with the domain $K \times Q$ and the following multiplication: $(k, q)(\ell, p) = (k \cdot (\varphi(q)(\ell)), qp)$, where $\cdot$ denotes the multiplication in K (note that $\varphi(q) \in$ Aut(K) and hence $\varphi(q)(\ell) \in K$).

The following result is stated in [116].

Theorem 4.8. *Let K and Q be finitely generated groups and let $\varphi : Q \to$ Aut(K) be a homomorphism. Then, for the semidirect product $K \rtimes_\varphi Q$ we have* WP($K \rtimes_\varphi Q$) $\leq_m^{\log}$ {WP(Q), CWP(K)}.

Proof. Let us consider a word $(k_1, q_1)(k_2, q_2) \cdots (k_n, q_n)$, where k_i (respectively, q_i) is a generator of K (respectively, Q). In $K \rtimes_\varphi Q$ we have

$$(k_1, q_1)(k_2, q_2) \cdots (k_n, q_n) = (\theta_1(k_1)\theta_2(k_2) \cdots \theta_n(k_n), q_1 q_2 \cdots q_n),$$

where $\theta_i \in$ Aut(K) is the automorphism defined by

$$\theta_i = \varphi(q_1 \cdots q_{i-1}) = \varphi(q_1) \cdots \varphi(q_{i-1})$$

for $1 \leq i \leq n$ (note that $\theta_1 = \mathsf{id}_K$). By Lemma 3.12, we can compute in logarithmic space an SLP $\mathbb{A}$ over the generators of K, which produces the word $\theta_1(k_1)\theta_2(k_2)\cdots\theta_n(k_n)$. We have $(k_1, q_1)(k_2, q_2)\cdots(k_n, q_n) = 1$ in $K \rtimes_\varphi Q$ if and only if $q_1 q_2 \cdots q_n = 1$ in Q and $\mathsf{val}(\mathbb{A}) = 1$ in K. This proves the proposition. $\square$

The semidirect product $G = K \rtimes_\varphi Q$ is a an extension of K by Q, i.e., K is a normal subgroup of G with quotient $G/K \simeq Q$. A reasonable generalization of Theorem 4.8 would be $\mathsf{WP}(G) \leq_m^{\log} (\mathsf{WP}(G/K), \mathsf{CWP}(K))$. But this cannot be true: There exist finitely generated groups G, Q, and K such that (i) $Q = G/K$, (ii) Q and K have computable word problems, and (iii) G has an undecidable word problem [14]. On the other hand, if we require additionally that Q is finitely presented (in fact, Q recursively presented suffices), then G must have a computable word problem [36]. For the special case that the quotient $Q = G/K$ has a polynomial time word search problem (see Definition 2.24), we can prove the following.

Theorem 4.9. *Let K be a finitely generated normal subgroup of G such that the quotient $Q = G/K$ is finitely presented and has a polynomial Dehn function and $\mathsf{WSP}(Q)$ can be solved in polynomial time. Then $\mathsf{WP}(G) \leq_m^P \mathsf{CWP}(K)$.*

Proof. Let Σ be a finite generating set for K and let $Q = G/K = \langle \Gamma \mid R \rangle$ with Γ a finite generating set for Q and $R \subseteq (\Gamma \cup \Gamma^{-1})^*$ a finite set of relators for Q. Let $\varphi : G \to Q$ be the canonical surjective homomorphism and choose a mapping $h : Q \to G$ with $h(1) = 1$ and $\varphi(h(a)) = a$ for $a \in Q$. The set $\Sigma \cup h(\Gamma)$ generates G and there exists a so-called factor set $f : Q \times Q \to K$ such that $h(a)h(b) = f(a, b)h(ab)$ for $a, b \in Q$. Note that in general h is not a homomorphism. $\square$

Let us take a word $w \in (\Sigma \cup \Sigma^{-1} \cup h(\Gamma \cup \Gamma^{-1}))^*$ and factorize w as

$$w = w_0 h(a_1) w_1 h(a_2) \cdots w_{n-1} h(a_n) w_n$$

with $w_i \in (\Sigma \cup \Sigma^{-1})^*$ $(0 \leq i \leq n)$ and $a_i \in \Gamma \cup \Gamma^{-1}$ $(1 \leq i \leq n)$.

The word $a_1 a_2 \cdots a_n \in (\Gamma \cup \Gamma^{-1})^*$ represents the group element $\varphi(w) \in Q$. In a first step, we check in polynomial time whether $\varphi(w) = 1$ in Q. Note that $\mathsf{WP}(Q)$ can be solved in polynomial time, since it reduces to $\mathsf{WSP}(Q)$.

If $\varphi(w) \neq 1$ in Q, we know that $w \neq 1$ in G. Hence, assume that $\varphi(w) = a_1 a_2 \cdots a_n = 1$ in Q for the rest of the proof. Hence, w represents an element of the normal subgroup K. Our goal will be to construct in polynomial time an SLP $\mathbb{C}$ over the alphabet $\Sigma \cup \Sigma^{-1}$ such that $w = \mathsf{val}(\mathbb{C})$ in G. Hence, $w = 1$ in G if and only if $\mathsf{val}(\mathbb{C}) = 1$ in K.

We first construct in polynomial time (using $\mathsf{WSP}(Q)$) a representation

$$a_1 a_2 \cdots a_n = \prod_{i=1}^{m} u_i r_i u_i^{-1} \text{ in the free group } F(\Gamma), \tag{4.1}$$

where $r_1, \ldots, r_m \in R \cup R^{-1}$ are relators or inverse relators for Q.

Consider the semi-Thue system

$$S = \{(aa^{-1}, \varepsilon), (\varepsilon, aa^{-1}) \mid a \in \Gamma \cup \Gamma^{-1}\} \cup \{(r, \varepsilon) \mid r \in R \cup R^{-1}\}.$$

For every rule $(v, w) \in S$, we have $\varphi(h(v)) = \varphi(h(w))$ in Q; here h is extended to a homomorphism $h : (\Gamma \cup \Gamma^{-1})^* \to G$. Hence, there exists a word $s_{v,w} \in (\Sigma \cup \Sigma^{-1})^*$ such that $h(v) = h(w)s_{v,w}$ in G. Let us fix these words $s_{v,w}$.

Using the identity (4.1), we can construct in polynomial time a derivation

$$a_1 a_2 \cdots a_n = u_0 \to_S u_1 \to_S u_2 \to_S \cdots \to_S u_k = \varepsilon \qquad (4.2)$$

of polynomial length in n.

Note that $h(a_1)h(a_2) \cdots h(a_n)$ represents an element of K in G. Using the derivation (4.2), we will compute in polynomial time an SLP $\mathbb{A}$ over $\Sigma \cup \Sigma^{-1}$ (the monoid generating set of the normal subgroup K) such that in the group G, we have

$$h(a_1)h(a_2) \cdots h(a_n) = \mathsf{val}(\mathbb{A}). \qquad (4.3)$$

For this, we will compute for every $0 \le i \le k$ an SLP $\mathbb{A}_i$ over $\Sigma \cup \Sigma^{-1}$ such that

$$h(a_1)h(a_2) \cdots h(a_n) = h(u_i)\mathsf{val}(\mathbb{A}_i)$$

in the group G. Clearly, for $\mathbb{A}_0$ we choose an SLP that generates the empty word. For the induction step, let us take $0 \le i < k$ and assume that we have an SLP $\mathbb{A}_i$ such that $h(a_1)h(a_2) \cdots h(a_n) = h(u_i)\mathsf{val}(\mathbb{A}_i)$ in the group G. Since $u_i \to_S u_{i+1}$ there exists a rule $(v, w) \in S$ and $x, y \in (\Gamma \cup \Gamma^{-1})^*$ such that $u_i = xvy$ and $u_{i+1} = xwy$. Since $h(v) = h(w)s_{v,w}$ in G, we get

$$
\begin{aligned}
h(u_i) &= h(x)h(v)h(y) \\
&= h(x)h(w)s_{v,w}h(y) \\
&= h(x)h(w)h(y)(h(y)^{-1}s_{v,w}h(y)) \\
&= h(u_{i+1})(h(y)^{-1}s_{v,w}h(y))
\end{aligned}
$$

in the group G. Since K is a normal subgroup of G, the word $h(y)^{-1}s_{v,w}h(y)$ can be certainly rewritten as a word over $\Sigma \cup \Sigma^{-1}$. At this point, we need the following claim:

Claim. Given words $h(b_1)h(b_2) \cdots h(b_j)$ with $b_1, \ldots, b_j \in \Gamma \cup \Gamma^{-1}$ and $c_1 c_2 \cdots c_\ell \in (\Sigma \cup \Sigma^{-1})^*$, we can construct in polynomial time an SLP $\mathbb{B}$ over $\Sigma \cup \Sigma^{-1}$ such that

$$h(b_j)^{-1} \cdots h(b_2)^{-1}h(b_1)^{-1}c_1 c_2 \cdots c_\ell h(b_1)h(b_2) \cdots h(b_j) = \mathsf{val}(\mathbb{B})$$

in the group G.

Proof of the claim. It suffices to construct in polynomial time SLPs $\mathbb{B}_i$ such that

$$h(b_j)^{-1} \cdots h(b_2)^{-1} h(b_1)^{-1} c_i h(b_1) h(b_2) \cdots h(b_j) = \mathsf{val}(\mathbb{B}_i)$$

in the group G. Then $\mathbb{B}$ can be defined as an SLP that generates the concatenation $\mathsf{val}(\mathbb{B}_1) \cdots \mathsf{val}(\mathbb{B}_\ell)$.

For $b \in \Gamma \cup \Gamma^{-1}$ let $\psi_b : K \to K$ be the automorphism of K defined by $\psi_b(x) = h(b)^{-1} x h(b)$. Thus, we have

$$h(b_j)^{-1} \cdots h(b_2)^{-1} h(b_1)^{-1} c_i h(b_1) h(b_2) \cdots h(b_j) = \psi_{b_j}(\cdots \psi_{b_2}(\psi_{b_1}(c_i)) \cdots)$$

An application of Lemma 3.12 therefore proves the claim.

Using our claim, we can construct in polynomial time an SLP $\mathbb{B}$ over $\Sigma \cup \Sigma^{-1}$ such that

$$h(y)^{-1} s_{v,w} h(y) = \mathsf{val}(\mathbb{B})$$

in the group G (note that the word y is a suffix of the word u_i that was constructed in polynomial time before). Hence, in the group G we have

$$\begin{aligned}
h(a_1) h(a_2) \cdots h(a_n) &= h(u_i) \mathsf{val}(\mathbb{A}_i) \\
&= h(u_{i+1})(h(y)^{-1} s_{v,w} h(y)) \mathsf{val}(\mathbb{A}_i) \\
&= h(u_{i+1}) \mathsf{val}(\mathbb{B}) \mathsf{val}(\mathbb{A}_i).
\end{aligned}$$

Hence, we can define $\mathbb{A}_{i+1}$ as an SLP that produces the concatenation $\mathsf{val}(\mathbb{B}) \mathsf{val}(\mathbb{A}_i)$.

Now, we can finish the proof of the theorem. We have

$$\begin{aligned}
w &= w_0 h(a_1) w_1 h(a_2) w_2 h(a_3) \cdots w_{n-1} h(a_n) w_n \\
&= w_0 \, (h(a_1) w_1 h(a_1)^{-1}) \, (h(a_1) h(a_2) w_2 h(a_2)^{-1} h(a_1)^{-1}) \cdots \\
&\qquad (h(a_1) \cdots h(a_n) w_n h(a_n)^{-1} \cdots h(a_1)^{-1}) \, (h(a_1) \cdots h(a_n))
\end{aligned}$$

in the group G. Using our claim, we can compute in polynomial time for every $1 \le i \le n$ an SLP $\mathbb{B}_i$ over $\Sigma \cup \Sigma^{-1}$ such that

$$w = w_0 \mathsf{val}(\mathbb{B}_1) \cdots \mathsf{val}(\mathbb{B}_n)(h(a_1) \cdots h(a_n)) \overset{(4.3)}{=} w_0 \mathsf{val}(\mathbb{B}_1) \cdots \mathsf{val}(\mathbb{B}_n) \mathsf{val}(\mathbb{A}),$$

where the SLP $\mathbb{A}$ over $\Sigma \cup \Sigma^{-1}$ was computed in polynomial time before. Hence, to check whether $w = 1$ in K, it suffices to check whether $w_0 \mathsf{val}(\mathbb{B}_1) \cdots \mathsf{val}(\mathbb{B}_n) \mathsf{val}(\mathbb{A}) = 1$ in the normal subgroup K. But this is an instance of $\mathsf{CWP}(K)$, since one can easily construct an SLP for $w_0 \mathsf{val}(\mathbb{B}_1) \cdots \mathsf{val}(\mathbb{B}_n) \mathsf{val}(\mathbb{A})$ from $w_0, \mathbb{B}_1, \ldots, \mathbb{B}_n$, and $\mathbb{A}$. This concludes the proof. $\qquad \square$

Theorem 4.9 is stated in [116] for the special case that Q is an automatic group.

4.3   The Compressed Word Problem in Finite Groups

The simplest finitely generated groups are the finite groups. So let us continue our investigations on the compressed word problem with finite groups.

Theorem 4.10. *For every finite group G, $\mathsf{CWP}(G)$ can be solved in polynomial time. Moreover, if G is a non-solvable finite group, then $\mathsf{CWP}(G)$ is $\mathbf{P}$-complete.*

Proof. The first statement is an immediate corollary of Theorem 3.11: Choose G as the generating set for G. The set of all words over G that evaluate to the identity in the finite group can be accepted by a finite automaton (with state set G). Hence, for a given SLP $\mathbb{A}$ over the alphabet G we can check in polynomial time whether $\mathsf{val}(\mathbb{A})$ evaluates to the identity of G.

The second statement is implicitly shown in [15] and is based on Barrington's proof of $\mathbf{NC}^1$-completeness of the word problem for a finite non-solvable group [11]. Let G be a finite non-solvable group. Hence, there exists a subgroup H such that $H = [H, H]$ ($[H, H]$ is the commutator subgroup of G, i.e., the subgroup generated by all commutators $aba^{-1}b^{-1}$ with $a, b \in G$). We can without loss of generality assume that $G = H$ and thus $G = [G, G]$. We choose the generating set $G \setminus \{1\}$ for G. We prove $\mathbf{P}$-hardness of $\mathsf{CWP}(G)$ by a reduction from circuit value problem; see Example 1.40. So, let $\mathscr{C}$ be a Boolean circuit. Since $\vee$ can be expressed with $\wedge$ and $\neg$, we can assume that $\mathscr{C}$ has no $\vee$-gates. We now construct an SLP $\mathbb{A}$ as follows: For every gate A of the circuit $\mathscr{C}$ and every nontrivial group element $g \in G \setminus \{1\}$, we introduce a variable A_g in the SLP $\mathbb{A}$. We will have $\mathsf{val}_{\mathbb{A}}(A_g) = 1$ in G if gate A evaluates to 0 and $\mathsf{val}_{\mathbb{A}}(A_g) = g$ in G if gate A evaluates to 1. If A is an input gate of the circuit $\mathscr{C}$, then we set $\mathsf{rhs}(A_g) = \varepsilon$ if gate A is labeled with 0 and $\mathsf{rhs}(A_g) = g$ if gate A is labeled with 1. Next, assume that A is a $\neg$-gate and let B be the unique input gate for A. We set $\mathsf{rhs}(A_g) = B_{g^{-1}}g$. Finally, let A be an $\wedge$-gate and let B and C be the two input gates for A. Since $G = [G, G]$, we can write g as a product of commutators $g = \prod_{i=1}^{n}(g_i^{-1}h_i^{-1}g_ih_i)$, where without loss of generality $n \leq |G|$ (which is a constant in our consideration) and $g_1, h_1, \dots, g_n, h_n \in G \setminus \{1\}$. We define

$$\mathsf{rhs}(A_g) = \prod_{i=1}^{n}(B_{g_i^{-1}}C_{h_i^{-1}}B_{g_i}C_{h_i}).$$

Note that if B evaluates to 0, then for all $1 \leq i \leq n$, $\mathsf{val}_{\mathbb{A}}(B_{g_i^{-1}}) = \mathsf{val}_{\mathbb{A}}(B_{g_i}) = 1$ in the group G. Hence, also

$$\mathsf{val}_{\mathbb{A}}(B_{g_i^{-1}})\mathsf{val}_{\mathbb{A}}(C_{h_i^{-1}})\mathsf{val}_{\mathbb{A}}(B_{g_i})\mathsf{val}_{\mathbb{A}}(C_{h_i}) = 1$$

in G and thus $\mathsf{val}_{\mathbb{A}}(A_g) = 1$ in G. If C evaluates to 0, we can argue analogously. On the other hand, if both B and C evaluate to 1, then for all $1 \leq i \leq n$,

$$\mathsf{val}_{\mathbb{A}}(B_{g_i^{-1}})\mathsf{val}_{\mathbb{A}}(C_{h_i^{-1}})\mathsf{val}_{\mathbb{A}}(B_{g_i})\mathsf{val}_{\mathbb{A}}(C_{h_i}) = g_i^{-1}h_i^{-1}g_ih_i$$

in G. Hence, $\mathsf{val}_{\mathbb{A}}(A_g) = g$ in G.

Finally, if O is the output gate of the circuit $\mathscr{C}$, then we add a new variable S to $\mathbb{A}$ (the start variable) and set $\mathrm{rhs}_\mathbb{A} = O_g g^{-1}$ for an arbitrary $g \in G \setminus \{1\}$. We get $\mathsf{val}(\mathbb{A}) = 1$ in G if and only if the circuit $\mathscr{C}$ evaluates to 1. It is not hard to show that the SLP $\mathbb{A}$ can be constructed with a logspace transducer from the circuit $\mathscr{C}$. $\square$

From Theorem 4.10 it follows that there exists a fixed regular language L for which it is **P**-complete to check whether $\mathsf{val}(\mathbb{A}) \in L$ for a given SLP $\mathbb{A}$ (see the remark after Theorem 3.11): Take any non-solvable finite group G (e.g., $G = A_5$). Then the set of all words over $G \setminus \{1\}$ whose product in G is 1 is a regular language.

For a finite solvable group, the compressed word problem is unlikely to be **P**-hard. Theorem 4.2 from [15] implies that the compressed word problem for a finite solvable group belongs to the complexity class **DET**, which is the class of all problems that are $\mathbf{NC}^1$-reducible to the problem of computing the determinant of a given integer matrix. It is known that $\mathbf{DET} \subseteq \mathbf{NC}^2$; see [162] for more details concerning these classes.

4.4 The Compressed Word Problem in Free Groups

In this section, we prove that the compressed word problem for a free group can be solved in polynomial time. This was shown in [110]. Later, we will prove more general results (see Sect. 5.1). Nevertheless, we decided to present details for the free group case since it shows the principal ideas for other more technical constructions in Sect. 5.1 and Chap. 6. Recall the definition of CSLPs (Definition 3.3).

Theorem 4.11. *The compressed word problem for a finitely generated free group belongs to* **P**.

Proof. Consider a free group $F(\Gamma)$ with Γ finite, and let $\mathbb{A} = (V, \Gamma \cup \Gamma^{-1}, S, \mathrm{rhs}_\mathbb{A})$ be an SLP in Chomsky normal form over the terminal alphabet $\Gamma \cup \Gamma^{-1}$. We will compute in polynomial time a CSLP $\mathbb{B} = (V, \Gamma \cup \Gamma^{-1}, S, \mathrm{rhs}_\mathbb{B})$ such that $\mathsf{val}(\mathbb{B}) = \mathsf{NF}(\mathsf{val}(\mathbb{A}))$ (recall the definition of the normal form mapping NF from Sect. 2.1). By Theorem 3.14 one can transform in polynomial time $\mathbb{B}$ into an equivalent SLP $\mathbb{B}'$. Then, $\mathsf{val}(\mathbb{A}) = 1$ in $F(\Gamma)$ if and only if $\mathsf{val}(\mathbb{B}') = \varepsilon$, which can be easily checked in polynomial time. The right-hand side mapping $\mathrm{rhs}_\mathbb{B}$ will be defined inductively over the hierarchical order of $\mathbb{A}$ in such a way that $\mathsf{val}_\mathbb{B}(A) = \mathsf{NF}(\mathsf{val}_\mathbb{A}(A))$ for every $A \in V$.

Consider a variable $A \in V$. If $\mathrm{rhs}_\mathbb{A}(A) = a \in \Gamma \cup \Gamma^{-1}$, then we set $\mathrm{rhs}_\mathbb{B}(A) = a$. Now assume that $\mathrm{rhs}_\mathbb{A}(A) = BC$ for variables $B, C \in V$. So, for the current CSLP $\mathbb{B}$ we already have $\mathsf{val}_\mathbb{B}(B) = \mathsf{NF}(\mathsf{val}_\mathbb{A}(B))$ and $\mathsf{val}_\mathbb{B}(C) = \mathsf{NF}(\mathsf{val}_\mathbb{A}(C))$. Using Theorem 3.14 for the current CSLP $\mathbb{B}$, we can construct in polynomial time two SLPs $\mathbb{B}_B$ and $\mathbb{B}_C$ such that $\mathsf{val}_\mathbb{B}(B) = \mathsf{val}(\mathbb{B}_B)$ and $\mathsf{val}_\mathbb{B}(C) = \mathsf{val}(\mathbb{B}_C)$.

Let $\mathsf{val}(\mathbb{B}_B) = u$ and $\mathsf{val}(\mathbb{B}_C) = v$. By Proposition 3.9(2) we can compute the lengths $m = |u|$ and $n = |v|$ in polynomial time. Let x be the longest word such that there exist u' and v' with $u = u'x$ and $v = x^{-1}v'$. Then, $\mathsf{NF}(\mathsf{val}_{\mathbb{A}}(A)) = u'v'$. If we can compute the length $k = |x|$ in polynomial time, then we can define $\mathsf{rhs}_{\mathbb{B}}(A) = B[1 : m - k]C[k + 1 : n]$.

Hence, it remains to compute the length of x in polynomial time. This can be done using a binary search approach. First, from the SLP $\mathbb{B}_B$ we can easily compute (by inverting all right-hand sides) an SLP $\mathbb{B}'_B$ such that $\mathsf{val}(\mathbb{B}'_B) = u^{-1}$. Then, it remains to compute the length of the longest common prefix of $\mathsf{val}(\mathbb{B}'_B)$ and $\mathsf{val}(\mathbb{B}_C)$. We first check in polynomial time whether $\mathsf{val}(\mathbb{B}'_B) = \mathsf{val}(\mathbb{B}_C)$ using Theorem 3.17. If this holds, then $|x| = m = n$. Now assume that $\mathsf{val}(\mathbb{B}'_B) \neq \mathsf{val}(\mathbb{B}_C)$. We can assume that $m = n$ is a power of 2. If, e.g., $m \leq n$ and 2^e is the smallest power of 2 that is at least n, then modify $\mathbb{B}'_B$ and $\mathbb{B}_C$ such that they generate words $u^{-1}b^{2^e - m}$ and $vb^{2^e - n}$ where $b \notin \Gamma \cup \Gamma^{-1}$ is a new symbol.

So, assume that $m = n = 2^e$. Note that $2^e \leq 2n$. We start our binary search with $\ell = p = 2^{e-1}$ and check in polynomial time (using Theorems 3.17 and 3.14) whether $\mathsf{val}(\mathbb{B}'_B)[1 : \ell] = \mathsf{val}(\mathbb{B}_C)[1 : \ell]$. If this is true, we set $p := p/2$ and $\ell := \ell + p$; otherwise set $p := p/2$ and $\ell := \ell - p$. We continue until $p = 1/2$, then ℓ is the length of x. The number of iterations is $e \in \mathcal{O}(\log n)$ and hence bounded by $\mathcal{O}(|\mathbb{A}|)$. This proves the theorem.   □

Schleimer proved a far-reaching generalization of Theorem 4.11.

Theorem 4.12 (cf. Schleimer, 2013, personal communication). *For every word hyperbolic group G, $\mathsf{CWP}(G)$ can be solved in polynomial time.*

In Sect. 5.1 we prove another generalization of Theorem 4.11: For every graph group G, $\mathsf{CWP}(G)$ can be solved in polynomial time (Corollary 5.6).

We can also show that the compressed word problem for a free group of rank at least 2 is **P**-complete.

Theorem 4.13. *The compressed word problem for a finitely generated free group of rank at least 2 is **P**-complete.*

Proof. It suffices to prove **P**-hardness for $F_2 = F(\{a, b\})$, which will be done by a reduction from the monotone circuit value problem; see Example 1.40. Let $\Gamma = \{a, b\}$.

Robinson has shown in [148, Theorem 6.3] that the (uncompressed) word problem for F_2 is $\mathbf{NC}^1$-hard. We will use the following facts from his proof: Let $x, y \in (\Gamma \cup \Gamma^{-1})^*$ such that $|x| = |y| = k$ and $|x|_a - |x|_{a^{-1}} = |y|_a - |y|_{a^{-1}} = 0$. Then, if we interpret x and y as elements from F_2, the following holds:

$$(x = 1) \vee (y = 1) \quad \Leftrightarrow \quad a^{-3k}xa^{3k}ya^{-3k}x^{-1}a^{3k}y^{-1} = 1$$

$$(x = 1) \wedge (y = 1) \quad \Leftrightarrow \quad a^{-3k}xa^{3k}ya^{-3k}xa^{3k}y = 1$$

Note that the words on the right of these equivalences have length $16k$ and that the number of a's minus the number of a^{-1}'s is again 0.

Now let $\mathscr{C}$ be a monotone Boolean circuit. Without loss of generality we can assume that $\mathscr{C}$ is layered, i.e., the gates of $\mathscr{C}$ are partitioned into n layers and a gate in layer $i > 1$ receives its inputs from layer $i - 1$; see, e.g., [67, Problem A.1.6]. Layer 1 contains the input gates and layer n contains the unique output gate. We now construct an SLP $\mathbb{A}$ as follows. For every gate z of $\mathscr{C}$, G contains two variables A_z and $A_{z^{-1}}$. We will have $\mathsf{val}_{\mathbb{A}}(A_z) = 1$ in F_2 if and only if gate z of the circuit evaluates to 1. The variable $A_{z^{-1}}$ evaluates to the inverse of $\mathsf{val}_{\mathbb{A}}(A_z)$ in F_2. Moreover, we will have $|\mathsf{val}_{\mathbb{A}}(A_z)| = |\mathsf{val}_{\mathbb{A}}(A_{z^{-1}})| = 2 \cdot 16^{i-1}$ if z is located in the i-th layer of the circuit ($1 \leq i \leq n$).

For every input gate x in layer 1 we define the right-hand sides of A_x and $A_{x^{-1}}$ as follows:

$$\mathrm{rhs}(A_x) = \begin{cases} aa^{-1} & \text{if gate } x \text{ is labeled with } 1 \\ b^2 & \text{if gate } x \text{ is labeled with } 0 \end{cases}$$

$$\mathrm{rhs}(A_{x^{-1}}) = \begin{cases} aa^{-1} & \text{if gate } x \text{ is labeled with } 1 \\ b^{-2} & \text{if gate } x \text{ is labeled with } 0 \end{cases}$$

If z is an $\vee$-gate in the i-th layer ($i \geq 2$) with input gates x and y from the $(i-1)$-th layer, then the right-hand sides for A_z and $A_{z^{-1}}$ are

$$\mathrm{rhs}(A_z) = a^{-6 \cdot 16^{i-2}} A_x a^{6 \cdot 16^{i-2}} A_y a^{-6 \cdot 16^{i-2}} A_{x^{-1}} a^{6 \cdot 16^{i-2}} A_{y^{-1}} \text{ and}$$

$$\mathrm{rhs}(A_{z^{-1}}) = A_y a^{-6 \cdot 16^{i-2}} A_x a^{6 \cdot 16^{i-2}} A_{y^{-1}} a^{-6 \cdot 16^{i-2}} A_{x^{-1}} a^{6 \cdot 16^{i-2}}.$$

Note that the binary codings of the exponents $6 \cdot 16^{i-2}$ have polynomial length and hence each of the above productions can be replaced by a sequence of ordinary productions. Moreover, if $|\mathsf{val}(A_u)| = 2 \cdot 16^{i-2}$ for $u \in \{x, x^{-1}, y, y^{-1}\}$ (which is true if x and y are located in the first layer, i.e., $i = 2$), then $|\mathsf{val}(A_z)| = |\mathsf{val}(A_{z^{-1}})| = 2 \cdot 16^{i-1}$. If z is an $\wedge$-gate in the i-th layer ($i \geq 2$) with input gates x and y, then the right-hand sides for A_z and $A_{z^{-1}}$ are

$$\mathrm{rhs}(A_z) = a^{-6 \cdot 16^{i-2}} A_x a^{6 \cdot 16^{i-2}} A_y a^{-6 \cdot 16^{i-2}} A_x a^{6 \cdot 16^{i-2}} A_y \text{ and}$$

$$\mathrm{rhs}(A_{z^{-1}}) = A_{y^{-1}} a^{-6 \cdot 16^{i-2}} A_{x^{-1}} a^{6 \cdot 16^{i-2}} A_{y^{-1}} a^{-6 \cdot 16^{i-2}} A_{x^{-1}} a^{6 \cdot 16^{i-2}}.$$

Once again, these definitions can be replaced by sequences of ordinary definitions. Let o be the unique output gate of the circuit $\mathscr{C}$. Then, by the result from [148], the circuit $\mathscr{C}$ evaluates to 1 if and only if $\mathsf{val}_{\mathbb{A}}(A_o) = 1$ in F_2. □

Let us conclude this section with a slight generalization of Theorem 4.11 to a uniform setting, where the rank of the free group is part of the input.

Theorem 4.14. *The following problem is P-complete:*

input: A finite alphabet Γ and an SLP $\mathbb{A}$ over the alphabet $\Gamma \cup \Gamma^{-1}$.
question: Does $\mathsf{val}(\mathbb{A}) = 1$ hold in $F(\Gamma)$?

Proof. By Theorem 4.13 it suffices to prove membership in **P**. Let $\Gamma = \{a_1, \ldots, a_n\}$. Then the mapping h with $h(a_i) = a^i b a^{-i}$ extends to an embedding of $F(\Gamma)$ into $F(a, b)$; see, e.g., [119, Proposition 3.1]. By Proposition 3.9(4) we can compute an SLP $\mathbb{B}$ for $h(\mathsf{val}(\mathbb{A}))$ from $\mathbb{A}$ and Γ in polynomial time. This proves the theorem by Theorem 4.11. $\Box$

4.5   The Compressed Word Problem for Finitely Generated Linear Groups

Recall the definition of the randomized complexity classes **RP** and **coRP** from Sect. 1.3.7. These classes are located between **P** and **NP**. Moreover, there is some evidence from complexity theory that **P** = **RP** = **coRP**. This makes the following result from [116] for finitely generated linear groups interesting.

Theorem 4.15. *Let G be a finitely generated linear group. Then, the compressed word problem for G belongs to* **coRP**.

Proof. Let G be linear over the field F. Assume first that F has characteristic 0. Recall the polynomial identity testing problem $\mathsf{PIT}(\mathbb{Z}) \in$ **coRP**; see Definition 1.48 and Theorem 1.49. Since **coRP** is closed under polynomial time many-one reductions, it suffices to show $\mathsf{CWP}(G) \leq_m^P \mathsf{PIT}(\mathbb{Z})$. Let Γ be a finite generating set of G and let $\mathbb{A} = (V, S, \mathrm{rhs})$ be an SLP in Chomsky normal form over $\Gamma \cup \Gamma^{-1}$. By Theorem 2.19 we can assume that G is a finitely generated group of $(d \times d)$-matrices over the field of fractions $\mathbb{Q}(x_1, \ldots, x_n)$. Let M_a be the matrix corresponding to generator $a \in \Gamma \cup \Gamma^{-1}$. These matrices are fixed; they are not part of the input. We can write M_a as $\frac{1}{p(x_1, \ldots, x_n)} N_a$, where N_a is a $(d \times d)$-matrix over $\mathbb{Z}[x_1, \ldots, x_n]$ and $p(x_1, \ldots, x_n)$ is the same polynomial over $\mathbb{Z}$ for all generators a. For $A \in V$ let n_A be the length of the word $\mathsf{val}_{\mathbb{A}}(A)$; it can be computed in polynomial time for every $A \in V$ by Proposition 3.9(2). From the SLP $\mathbb{A}$ we can now build in polynomial time arithmetic circuits $\mathscr{C}_{i,j}$ ($1 \leq i, j \leq d$) in the variables $x_1, \ldots, x_n$ such that $\mathsf{val}(\mathscr{C}_{i,j}) = 0$ for all $1 \leq i, j \leq d$ if and only if $\mathsf{val}(\mathbb{A}) = 1$ in G. For this, we associate d^2 defined circuit variables $A_{i,j}$ ($1 \leq i, j \leq d$) with every SLP-variable $A \in V$. The circuit variable $A_{i,j}$ will evaluate to entry (i, j) of the matrix $p(x_1, \ldots, x_n)^{n_A} \mathsf{val}_{\mathbb{A}}(A)$. If $\mathrm{rhs}(A) = a \in \Gamma \cup \Gamma^{-1}$, then $\mathrm{rhs}(A_{i,j})$ is set to entry (i, j) of the matrix N_a. If $\mathrm{rhs}(A) = BC$, then we set

$$\mathrm{rhs}(A_{i,j}) = \sum_{k=1}^{d} B_{i,k} C_{k,j}.$$

Finally, we set the output variable of the circuit $\mathscr{C}_{i,j}$ to $S_{i,j}$ for $i \neq j$ and to $S_{i,j} - p(x_1, \ldots, x_n)^{n_S}$ for $i = j$.

From the circuits $\mathscr{C}_{i,j}$ we can easily construct an arithmetic circuit for the polynomial

$$p(x_1, \ldots, x_n, y, z) = \sum_{1 \le i,j \le d} y^i z^j \text{val}(\mathscr{C}_{i,j}).$$

Then $p(x_1, \ldots, x_n, y, z)$ is the zero polynomial if and only if every polynomial $\text{val}(\mathscr{C}_{i,j})$ is the zero polynomial.

The same arguments apply if F has prime characteristic p. In that case (using again Theorem 2.19) we can show $\text{CWP}(G) \le_m^P \text{PIT}(\mathbb{Z}_p)$. By Theorem 1.49, also polynomial identity testing over the coefficient ring $\mathbb{Z}_p$ belongs to **coRP**. $\square$

Examples of finitely generated linear groups are finitely generated polycyclic groups (which include finitely generated nilpotent groups), Coxeter groups, braid groups, and graph groups. Hence, for all these groups the compressed word problem belongs to **coRP**. The same holds for finitely generated metabelian groups since they embed into finite direct products of linear groups [164]. For finitely generated nilpotent groups, Coxeter groups, and graph groups, we will show that the compressed word problem even belongs to **P** (Theorem 4.19, Corollaries 5.6, and 5.7). In the next section, we will present a concrete group for which the compressed word problem is equivalent (with respect to polynomial time many-one reductions) to polynomial identity testing over the ring $\mathbb{Z}$.

4.6 The Compressed Word Problem for $SL_3(\mathbb{Z})$

Recall that the *special linear group* $SL_d(\mathbb{Z})$ is the group of all $(d \times d)$-matrices over $\mathbb{Z}$ with determinant 1. The group $SL_2(\mathbb{Z})$ is virtually free (has a free subgroup of finite index). Hence, by Theorem 4.4 and 4.11, its compressed word problem can be solved in polynomial time. For $SL_3(\mathbb{Z})$, the complexity of the compressed word problem exactly coincides with the complexity of polynomial identity testing over $\mathbb{Z}$, which is not known to be in **P**.

Theorem 4.16. $\text{CWP}(SL_3(\mathbb{Z}))$ *and* $\text{PIT}(\mathbb{Z})$ *are polynomial time many-one reducible to each other.*

Proof. Since $SL_3(\mathbb{Z})$ is linear over the field $\mathbb{Q}$ of characteristic zero, the proof of Theorem 4.15 shows that $\text{CWP}(SL_3(\mathbb{Z})) \le_m^P \text{PIT}(\mathbb{Z})$.

The proof for $\text{PIT}(\mathbb{Z}) \le_m^P \text{CWP}(SL_3(\mathbb{Z}))$ is based on a construction from Ben-Or and Cleve [16], which can be seen as an arithmetic version of Barrington's construction (that we used in the proof of Theorem 4.10). By Theorem 1.51 it suffices to construct in polynomial time from a variable-free arithmetic circuit $\mathscr{C} = (V, S, \text{rhs})$ over $\mathbb{Z}$ an SLP $\mathbb{A}$ over generators of $SL_3(\mathbb{Z})$ such that $\text{val}(\mathscr{C}) = 0$ if and only if $\text{val}(\mathbb{A})$ evaluates to the identity matrix.

The SLP $\mathbb{A}$ contains for every gate $A \in V$ and all $b \in \{-1, 1\}$ and $1 \le i, j \le 3$ with $i \neq j$ a variable $A_{i,j,b}$. Let us denote the matrix to which $A_{i,j,b}$ evaluates with $A_{i,j,b}$ as well. The SLP $\mathbb{A}$ is constructed in such a way that for every column vector $(x_1, x_2, x_3)^T \in \mathbb{Z}^3$, the following holds for the vector $(y_1, y_2, y_3)^T = A_{i,j,b}(x_1, x_2, x_3)^T$: $y_i = x_i + b \cdot \mathsf{val}_{\mathscr{C}}(A) \cdot x_j$ and $y_k = x_k$ for $k \in \{1, 2, 3\} \setminus \{j\}$.

Consider a gate A of the variable-free arithmetic circuit $\mathscr{C}$. Without loss of generality assume that $\mathscr{C}$ is in normal form. We make a case distinction on the right-hand side of A.

Case 1. $\mathsf{rhs}(A) = c \in \{-1, 1\}$. Then we set $\mathsf{rhs}(A_{i,j,b}) = \mathsf{Id}_3 + M_{i,j,b\cdot c}$, where Id_3 is the (3×3) identity matrix and all entries in the matrix $M_{i,j,b\cdot c}$ are 0, except for the entry at position (i, j), which is $b \cdot c$.

Case 2. $\mathsf{rhs}(A) = B + C$. Then we set $\mathsf{rhs}(A_{i,j,b}) = B_{i,j,b} C_{i,j,b}$.

Case 3. $\mathsf{rhs}(A) = B \cdot C$. Let $\{k\} = \{1, 2, 3\} \setminus \{i, j\}$. Then we set

$$\mathsf{rhs}(A_{i,j,1}) = B_{k,j,-1} C_{i,k,1} B_{k,j,1} C_{i,k,-1}$$

$$\mathsf{rhs}(A_{i,j,-1}) = B_{k,j,-1} C_{i,k,-1} B_{k,j,1} C_{i,k,1}$$

If $(y_1, y_2, y_3)^T = A_{i,j,1}(x_1, x_2, x_3)^T = B_{k,j,-1} C_{i,k,1} B_{k,j,1} C_{i,k,-1}(x_1, x_2, x_3)^T$, then we get $y_j = x_j$, $y_k = x_k + \mathsf{val}_{\mathscr{C}}(B) \cdot x_j - \mathsf{val}_{\mathscr{C}}(B) \cdot x_j = x_k$, and

$$y_i = x_i - \mathsf{val}_{\mathscr{C}}(C) \cdot x_k + \mathsf{val}_{\mathscr{C}}(C) \cdot (x_k + \mathsf{val}_{\mathscr{C}}(B) \cdot x_j) = x_i + \mathsf{val}_{\mathscr{C}}(C) \cdot \mathsf{val}_{\mathscr{C}}(B) \cdot x_j.$$

Similarly, if $(y_1, y_2, y_3)^T = A_{i,j,-1}(x_1, x_2, x_3)^T = B_{k,j,-1} C_{i,k,-1} B_{k,j,1} C_{i,k,1}(x_1, x_2, x_3)^T$, then we get $y_j = x_j$, $y_k = x_k + \mathsf{val}_{\mathscr{C}}(B) \cdot x_j - \mathsf{val}_{\mathscr{C}}(B) \cdot x_j = x_k$, and

$$y_i = x_i + \mathsf{val}_{\mathscr{C}}(C) \cdot x_k - \mathsf{val}_{\mathscr{C}}(C) \cdot (x_k + \mathsf{val}_{\mathscr{C}}(B) \cdot x_j) = x_i - \mathsf{val}_{\mathscr{C}}(C) \cdot \mathsf{val}_{\mathscr{C}}(B) \cdot x_j.$$

Finally, let $S_{1,2,1}$ be the start variable of $\mathbb{A}$. Then, we have $\mathsf{val}(\mathscr{C}) = 0$ if and only if for all $(x_1, x_2, x_3) \in \mathbb{Z}^3$ we have $S_{1,2,1}(x_1, x_2, x_3)^T = (x_1 + \mathsf{val}(\mathscr{C}) \cdot x_2, x_2, x_3) = (x_1, x_2, x_3)$ if and only if $\mathsf{val}(\mathbb{A})$ evaluates to Id_3. This proves the theorem. $\square$

4.7 The Compressed Word Problem for Finitely Generated Nilpotent Groups

Let us first recall the definition of a nilpotent group.

Definition 4.17 (lower central series, nilpotent group). The *lower central series* of the group G is the sequence of subgroups $G = G_1 \geq G_2 \geq G_3 \geq \cdots$ where $G_{i+1} = [G_i, G]$ (which is the subgroup of G_i generated by all commutators $g^{-1}h^{-1}gh$ for $g \in G_i$ and $h \in G$; by induction one can show that indeed $G_{i+1} \leq G_i$). The group G is nilpotent if there exists $i \geq 1$ with $G_i = 1$.

Robinson [148] has shown that the word problem for a finitely generated linear group belongs to the circuit complexity class $\mathbf{TC}^0 \subseteq \mathbf{L}$. Moreover, every nilpotent group is linear; see, e.g., [95]. Hence, by Theorem 4.15 the compressed word problem for a finitely generated nilpotent group belongs to $\mathbf{coRP}$. This upper bound was improved in [74] to $\mathbf{P}$. Our proof of this result uses a fact about unitriangular matrices. Recall that a $(d \times d)$-matrix $M = (a_{i,j})_{1 \leq i,j \leq d}$ over $\mathbb{Z}$ is unitriangular if $a_{i,i} = 1$ for all $1 \leq i \leq d$ and $a_{i,j} = 0$ for $i > j$, i.e., all entries below the diagonal are zero. For a matrix $M = (a_{i,j})_{1 \leq i,j \leq d}$ over $\mathbb{Z}$ let $|M| = \sum_{1 \leq i,j \leq d} |a_{i,j}|$. It is straightforward to show that $|M_1 \cdot M_2| \leq d^2 |M_1| \cdot |M_2|$ for two $(d \times d)$-matrices M_1 and M_2; see [108, Lemma 3]. Hence, for a product of m matrices we have

$$|M_1 \cdot M_2 \cdots M_m| \leq d^{2(m-1)} |M_1| \cdot |M_2| \cdots |M_n|.$$

The estimate in the following proposition is very rough but sufficient for our purpose.

Proposition 4.18. *Let $M_1, \ldots, M_m$ be a unitriangular $(d \times d)$-matrices over $\mathbb{Z}$ with $m \geq 2d$ and let $n = \max\{|M_i| \mid 1 \leq i \leq m\}$. For the product of these matrices we have*

$$|M_1 M_2 \cdots M_m| \leq d + (d-1) \binom{m}{d-1} d^{2(d-2)} n^{d-1}.$$

Proof. Let $A_i = M_i - \mathsf{Id}_d$; this is a matrix which has only zeros on the diagonal and below. Hence any product of at least d many matrices A_i is zero. We get

$$M_1 M_2 \cdots M_m = \prod_{i=1}^{m} (A_i + \mathsf{Id}_d) = \sum_{I \subseteq \{1,\ldots,m\}, |I| < d} \prod_{i \in I} A_i.$$

We have $|A_i| \leq n$ and hence

$$\left| \prod_{i \in I} A_i \right| \leq d^{2(|I|-1)} \prod_{i \in I} |A_i| \leq d^{2(|I|-1)} n^{|I|}$$

for $I \neq \emptyset$. Finally, we get

$$|M_1 M_2 \cdots M_m| \leq d + \sum_{i=1}^{d-1} \binom{m}{i} d^{2(i-1)} n^i$$

(the summand d is due to $|\mathsf{Id}_d| = d$). Since we assume $m \geq 2d$ we have $\binom{m}{i} \leq \binom{m}{d-1}$ for all $1 \leq i \leq d-1$. Hence, we get

$$|M_1 M_2 \cdots M_m| \leq d + (d-1) \binom{m}{d-1} d^{2(d-2)} n^{d-1}.$$

$\square$

Theorem 4.19. *Let G be a finitely generated nilpotent group. Then* $\mathsf{CWP}(G)$ *can be solved in polynomial time.*

Proof. Let G be a finitely generated nilpotent group. Then G has a finitely generated torsion-free nilpotent subgroup H such that the index $[G : H]$ is finite [95, Theorem 17.2.2]. By Theorem 4.4, it suffices to solve $\mathsf{CWP}(H)$ in polynomial time. Let Γ be a finite generating set for H. There exists $d \geq 1$ such that the finitely generated torsion-free nilpotent group H can be embedded into the group $\mathsf{UT}_d(\mathbb{Z})$ of unitriangular $(d \times d)$-matrices over $\mathbb{Z}$ [95, Theorem 17.2.5]. Let $\varphi : H \to \mathsf{UT}_d(\mathbb{Z})$ be this embedding. Let $n = \max\{|\varphi(a)| \mid a \in \Gamma \cup \Gamma^{-1}\}$. Note that n and d are constants in our consideration.

If the word $w \in (\Gamma \cup \Gamma^{-1})^*$ is given by an SLP $\mathbb{A} = (V, \Gamma \cup \Gamma^{-1}, S, \mathrm{rhs})$ in Chomsky normal form of size m, we can evaluate the SLP bottom-up in the group $\mathsf{UT}_d(\mathbb{Z})$ as follows: For every variable $A \in V$ we compute the matrix $\varphi(\mathsf{val}(A))$. If $\mathrm{rhs}(A) = BC$ and the matrices $\varphi(\mathsf{val}(B))$ and $\varphi(\mathsf{val}(C))$ are already computed, then $\varphi(\mathsf{val}(A))$ is set to the product of these two matrices. Since $|\mathsf{val}_{\mathbb{A}}(A)| \leq |w| \leq 2^m$, Proposition 4.18 implies

$$|\varphi(\mathsf{val}(A))| \leq d + (d-1)\binom{2^m}{d-1}d^{2(d-2)}n^{d-1} \in 2^{\mathscr{O}(m)}.$$

Hence, every entry in the matrix $\varphi(\mathsf{val}(A))$ can be represented with $\mathscr{O}(m)$ bits. Therefore, the evaluation can be accomplished in polynomial time. $\square$

By [9], the automorphism group of a finitely generated nilpotent group is finitely generated (even finitely presented) and hence, by Theorem 4.19 has a polynomial time word problem.

4.8   Wreath Products: Easy Word Problem but Difficult Compressed Word Problem

In this section, we will present a group G, for which the word problem can be solved in logarithmic space, but the compressed word problem is **coNP**-hard and thus not solvable in polynomial time unless $\mathbf{P} = \mathbf{NP}$ (which is equivalent to $\mathbf{P} = \mathbf{coNP}$). We start with the definition of a wreath product of two groups.

Definition 4.20 (wreath product). Let G and H be groups. Consider the direct sum

$$K = \bigoplus_{g \in G} H_g,$$

where H_g is a copy of H. We view K as the set

$$H^{(G)} = \{f : G \to H \mid f(g) \neq 1 \text{ for only finitely many } g \in G\}$$

of all mappings from G to H with finite support together with pointwise multiplication as the group operation, i.e., $(f_1 f_2)(g) = f_1(g) f_2(g)$. The group G has a natural left action on $H^{(G)}$ given by

$$gf(a) = f(g^{-1}a)$$

where $f \in H^{(G)}$ and $g, a \in G$. The corresponding homomorphism $\varphi : G \to \text{Aut}(H^{(G)})$ is defined by $\varphi(g)(f) = gf$ for $g \in G$ and $f \in H^{(G)}$. The corresponding semidirect product $H^{(G)} \rtimes_\varphi G$ is the *wreath product* $H \wr G$.[1] In other words:

- Elements of $H \wr G$ are pairs (f, g), where $g \in G$ and $f \in H^{(G)}$.
- The multiplication in $H \wr G$ is defined as follows: Let $(f_1, g_1), (f_2, g_2) \in H \wr G$. Then $(f_1, g_1)(f_2, g_2) = (f, g_1 g_2)$, where $f(a) = f_1(a) f_2(g_1^{-1} a)$.

The following intuition might be helpful: An element $(f, g) \in H \wr G$ can be thought of as a finite multiset of elements of $H \setminus \{1\}$ that are sitting at certain elements of G (the mapping f) together with the distinguished element $g \in G$, which can be thought of as a cursor moving in G. If we want to compute the product $(f_1, g_1)(f_2, g_2)$, we do this as follows: First, we shift the finite collection of H-elements that corresponds to the mapping f_2 by g_1 (the result is the mapping $g_1 f_2$): If the element $h \in H \setminus \{1\}$ is sitting at $a \in G$ (i.e., $f_2(a) = h$), then we remove h from a and put it to the new location $g_1 a \in H$. This new collection corresponds to the mapping $f_2' : a \mapsto f_2(g_1^{-1} a)$. After this shift, we multiply the two collections of H-elements pointwise: If in $a \in G$ the elements h_1 and h_2 are sitting (i.e., $f_1(a) = h_1$ and $f_2'(a) = h_2$), then we put the product $h_1 h_2$ into the location a. Finally, the new distinguished G-element (the new cursor position) becomes $g_1 g_2$.

Assume that $H = \langle \Sigma \mid R_H \rangle$ and $G = \langle \Gamma \mid R_G \rangle$ with $\Sigma \cap \Gamma = \emptyset$. Then we have

$$H \wr G \cong \langle \Sigma \cup \Gamma \mid R_G \cup R_H \cup \{[waw^{-1}, b] \mid a, b \in \Sigma, w \in (\Gamma \cup \Gamma^{-1})^*, w \neq 1 \text{ in } G\}\rangle;$$

see, e.g., [125]. In terms of the above intuition, the relator $[waw^{-1}, b]$ (recall that $[x, y] = xyx^{-1}y^{-1}$) expresses that the following two actions have the same effect:

- (i) Moving the cursor to (the G-element represented by) w, (ii) multiplying the H-element at the new cursor position with a, (iii) moving the cursor back to the origin, and (iv) finally multiplying the H-element at the origin with b.

---

[1]This wreath product is also called the restricted wreath product since only finitely supported mappings from G to H are considered and not all mappings.

- (i) Multiplying the H-element at the current cursor with b, (ii) moving the cursor to w, (iii) multiplying the H-element at the new cursor position with a, and finally (iv) moving the cursor back to the origin.

If G and H are finitely generated, then also $H \wr G$ is finitely generated. On the other hand, $H \wr G$ is finitely presented if and only if one of the following two cases holds: (i) $H = 1$ and G is finitely presented or (ii) H is finitely presented and G is finite; see [12]. The complexity of the word problem for wreath products was studied in [163].

Theorem 4.21. *If G is finitely generated non-abelian and $\mathbb{Z} \leq H$, then* $\mathsf{CWP}(G \wr H)$ *is* **coNP**-*hard.*

Proof. It suffices to prove the theorem for the wreath product $G \wr \mathbb{Z}$. Let $g, h \in G$ such that $gh \neq hg$ and let $\mathbb{Z}$ be generated by t. We prove the theorem by a reduction from the problem from Theorem 3.13. Hence, let $\mathbb{A}$ and $\mathbb{B}$ be SLPs over the alphabet $\{a, b\}$. Let $n = |\mathsf{val}(\mathbb{A})|$ and $m = |\mathsf{val}(\mathbb{B})|$. Moreover, let $\mathbb{A}_1$ (respectively, $\mathbb{B}_1$) be the SLP that results from $\mathbb{A}$ by replacing every occurrence of the symbol a by the $\mathbb{Z}$-generator t and every occurrence of the symbol b by tg (respectively, th). Similarly, let $\mathbb{A}_2$ (respectively, $\mathbb{B}_2$) be the SLP that results from $\mathbb{A}$ by replacing every occurrence of the symbol a by t and every occurrence of the symbol b by tg^{-1} (respectively, th^{-1}). From these SLPs it is easy to construct in logspace an SLP $\mathbb{C}$ such that

$$\mathsf{val}(\mathbb{C}) = \mathsf{val}(\mathbb{A}_1)t^{-n}\mathsf{val}(\mathbb{B}_1)t^{-m}\mathsf{val}(\mathbb{A}_2)t^{-n}\mathsf{val}(\mathbb{B}_2)t^{-m}$$

(SLPs for t^{-n} and t^{-m} can be obtained in logspace by replacing in $\mathbb{A}$ and $\mathbb{B}$, respectively, all occurrences of a and b by t^{-1}). Then, if there is no position $i \in \mathbb{N}$ such that $\mathsf{val}(\mathbb{A})[i] = \mathsf{val}(\mathbb{B})[i] = b$, we have $\mathsf{val}(\mathbb{C}) = 1$ in $G \wr \mathbb{Z}$. On the other hand, if there is a position $i \in \mathbb{N}$ such that $\mathsf{val}(\mathbb{A})[i] = \mathsf{val}(\mathbb{B})[i] = b$, then, if $(f, 0) \in G \wr \mathbb{Z}$ is the element represented by the word $\mathsf{val}(\mathbb{C})$, we have $f(i) = ghg^{-1}h^{-1} \neq 1$. Hence $\mathsf{val}(\mathbb{C}) \neq 1$ in $G \wr \mathbb{Z}$. $\square$

If G is a finite group, then the word problem for $G \wr \mathbb{Z}$ can be solved in logspace. This follows from [163], where it has been shown that if G_1 and G_2 are groups for which the word problem belongs to the circuit complexity class (uniform) $\mathbf{NC}^1$ (and this is the case for finite groups and $\mathbb{Z}$), then also the word problem for $G_1 \wr G_2$ belongs to $\mathbf{NC}^1 \subseteq \mathbf{L}$. Hence, for a finite non-abelian group G, the word problem for $G \wr \mathbb{Z}$ belongs to $\mathbf{L}$, but the compressed word problem is **coNP**-hard. Another group with an easy word problem but difficult compressed word problem is *Thompson's group*

$$F = \langle x_0, x_1, x_2, \ldots \mid x_n x_k = x_k x_{n+1} \text{ for all } k < n \rangle.$$

This group is actually finitely presented: $F = \langle a, b \mid [ab^{-1}, a^{-1}ba], [ab^{-1}, a^{-2}ba^2] \rangle$. The group F has several other nice representations, e.g., by piecewise linear homeomorphisms of the unit interval $[0, 1]$ or by certain tree diagrams; see [26] for more details. Thompson's group F is a subgroup of Thompson's group V, for which the word problem belongs to the circuit complexity class $\mathbf{AC}^1$ [18], which satisfies $\mathbf{NL} \subseteq \mathbf{AC}^1 \subseteq \mathbf{NC}^2 \subseteq \mathbf{P}$.

Theorem 4.22. $\mathsf{CWP}(F)$ *is* **coNP**-*hard.*

Proof. It is known that the wreath product $F \wr \mathbb{Z}$ is a subgroup of F [69]. Since F is non-abelian, the theorem follows from Theorem 4.21.   □

If G and H are finitely generated abelian groups, then the wreath product $H \wr G$ is metabelian ($H^{(G)}$ is a normal subgroup of $H \wr G$ with $(H \wr G)/H^{(G)} = G$). Hence, as remarked earlier, $H \wr G$ can be embedded into a direct product of linear groups and $\mathsf{CWP}(H \wr G)$ belongs to **coRP**. This fact can be slightly extended.

Corollary 4.23. *For every finitely generated abelian group H and every finitely generated virtually abelian group H (i.e., H is a finite extension of a finitely generated abelian group), $\mathsf{CWP}(H \wr G)$ belongs to* **coRP***.*

Proof. Assume that K is a finitely generated subgroup of index m in G. Then $H^m \wr K$ is isomorphic to a subgroup of index m in $H \wr G$ (see, e.g., [118]). If H is finitely generated abelian, then H^m is finitely generated abelian too and therefore $\mathsf{CWP}(H^m \wr K)$ belongs to **coRP**. Finally, we can apply Theorem 4.4 and the fact that **coRP** is closed under $\leq_m^P$.   □

It is well known that if N is a normal subgroup of G, then also $[N, N]$ is a normal subgroup of G. Hence, one can consider the quotient group $G/[N, N]$. The following result of Magnus [122] has many applications in combinatorial group theory.

Theorem 4.24 (Magnus embedding theorem). *Let F_k be a free group of rank k and let N be a normal subgroup of F_k. Then $F_k/[N, N]$ embeds into the wreath product $\mathbb{Z}^k \wr (F_k/N)$.*

We can use the Magnus embedding theorem to get

Theorem 4.25. *Let F_k be a free group of rank k and let N be a normal subgroup of F_k such that F_k/N is finitely generated virtually abelian. Then $\mathsf{CWP}(F_k/[N, N])$ belongs to* **coRP***.*

Proof. By the Magnus embedding theorem, the group $F_k/[N, N]$ embeds into the wreath product $\mathbb{Z}^k \wr (F_k/N)$. For the latter group, the compressed word problem belongs to **coRP** by Corollary 4.23.   □

Chapter 5
The Compressed Word Problem in Graph Products

In this chapter we will introduce an important operation in combinatorial group theory: graph products. A graph product is specified by a finite undirected graph, where every node is labeled with a group. The graph product specified by this group-labeled graph is obtained by taking the free product of all groups appearing in the graph, but elements from adjacent groups are allowed to commute. This operation generalizes free products as well as direct products. Graph groups were introduced by Green in her thesis [66]. Further results for graph products can be found in [51,75,83,129].

The main result of this chapter states that there is a polynomial time Turing-reduction from the compressed word problem for a graph product to the compressed word problems of the groups labeling the nodes of the graph. The material of this chapter is taken from [74].

5.1 Graph Products

In this section, we formally define graph products and state the main result of this chapter.

Definition 5.1 ((in)dependence alphabet). An independence alphabet is a pair (A, I), where A is an arbitrary set and $I \subseteq A \times A$ is an irreflexive and symmetric relation on A. The *dependence alphabet* associated with (A, I) is (A, D), where $D = (A \times A) \setminus I$. Note that the relation D is reflexive.

Let us fix for this subsection a *finite* independence alphabet (W, E) with $W = \{1, \ldots, n\}$ and finitely generated groups G_i for $i \in \{1, \ldots, n\}$. Let $G_i = \langle \Gamma_i \mid R_i \rangle$ with $\Gamma_i \cap \Gamma_j = \emptyset$ for $i \neq j$.

M. Lohrey, *The Compressed Word Problem for Groups*, SpringerBriefs in Mathematics,
DOI 10.1007/978-1-4939-0748-9__5, © Markus Lohrey 2014

Definition 5.2 (graph product). The *graph product* defined by $(W, E, (G_i)_{i \in W})$ is the following group:

$$\mathbb{G}(W, E, (G_i)_{i \in W}) = \left\langle \bigcup_{i=1}^{n} \Gamma_i \mid \bigcup_{i=1}^{n} R_i \cup \bigcup_{(i,j) \in E} \{[a, b] \mid a \in \Sigma_i, b \in \Sigma_j\} \right\rangle$$

In other words, we take the free product of all the groups $G_1, \ldots, G_n$, but elements $x \in G_i$, $y \in G_j$ with $(i, j) \in E$ are allowed to commute.

Clearly, if every G_i is finitely generated (respectively, finitely presented), then the same holds for $\mathbb{G}(W, E, (G_i)_{i \in W})$. If $E = \emptyset$, then $\mathbb{G}(W, E, (G_i)_{i \in W})$ is the free product $\mathbb{G}_1 * \mathbb{G}_2 * \cdots * \mathbb{G}_n$, and if (W, E) is a complete graph, then $\mathbb{G}(W, E, (G_i)_{i \in W})$ is the direct product $\prod_{i=1}^{n} G_i$. In this sense, the graph product construction generalizes free and direct products.

Note that graph groups (see Example 2.5) are exactly the graph products of copies of $\mathbb{Z}$. Graph products of copies of $\mathbb{Z}/2\mathbb{Z}$ are known as *right-angled Coxeter groups*; see [56] for more details.

Recently, it was shown that the word problem for a graph product of groups with logspace word problem can be solved in logspace too [50].

Theorem 5.3. *Let (W, E) be a fixed finite independence alphabet and for every $i \in W$ let G_i be a finitely generated group. If $\mathsf{WP}(G_i) \in \mathbf{L}$ for all $i \in W$, then also $\mathsf{WP}(\mathbb{G}(W, E, (G_i)_{i \in W})) \in \mathbf{L}$.*

Recall the definition of polynomial time Turing-reducibility $\leq_T^P$ (Definition 1.29). The main result of this chapter is

Theorem 5.4. *Let (W, E) be a finite independence alphabet and for every $i \in W$ let G_i be a finitely generated group. Then we have*

$$\mathsf{CWP}(\mathbb{G}(W, E, (G_i)_{i \in W})) \leq_T^P \{\mathsf{CWP}(G_i) \mid i \in W\}.$$

By taking $(W, E) = (\{1, 2\}, \emptyset)$, we get

Corollary 5.5. *Let G_1 and G_2 be finitely generated groups. Then $\mathsf{CWP}(G_1 * G_2) \leq_T^P \{\mathsf{CWP}(G_1), \mathsf{CWP}(G_2)\}$.*

By taking $G_I = \mathbb{Z}$ (respectively, $\mathbb{Z}/2\mathbb{Z}$) for every $i \in W$, we get

Corollary 5.6. *For every graph group G (respectively, right-angled Coxeter group G), $\mathsf{CWP}(G)$ belongs to $\mathbf{P}$.*

Building on results from [154], Laurence has shown in [102] that automorphism groups of graph groups are finitely generated. Recently, Day [44] proved that automorphism groups of graph groups are in fact finitely presented. Further structural results on automorphism groups of graph groups can be found in [33, 34]. Generalizing the main result from [102], it was shown in [43] that the automorphism group of a graph product of finitely generated Abelian groups is finitely generated. In particular, the automorphism group of a right-angled Coxeter group is finitely generated. From Corollary 5.6 and Theorem 4.6 it follows that the word problem for the automorphism group of a graph group or a right-angled Coxeter group can be solved in polynomial time. More generally, Theorem 4.6 and 5.4 yield a polynomial time algorithm for the word problem of a finitely generated subgroup of $\mathsf{Aut}(\mathbb{G}(W, E, (G_i)_{i \in W}))$, where every vertex group G_i has a polynomial time compressed word problem. It is not clear whether the full group $\mathsf{Aut}(\mathbb{G}(W, E, (G_i)_{i \in W}))$ is finitely generated in case every group $\mathsf{Aut}(G_i)$ is finitely generated.

Proposition 4.3, Theorem 4.4, and Corollary 5.6 imply that for every finite extension G of a subgroup of a graph group (one also says that G virtually embeds into a graph group), $\mathsf{CWP}(G)$ belongs to $\mathbf{P}$. Recently, this class of groups turned out to be very rich. It contains the following classes of groups:

- Coxeter groups (not only right-angled ones) [72]
- one-relator groups with torsion [165]
- fully residually free groups [165]
- fundamental groups of hyperbolic 3-manifolds [2]

Hence, we can state the following corollary.

Corollary 5.7. *For every group from one of the following classes the compressed word problem belongs to* $\mathbf{P}$*: Coxeter groups, one-relator groups with torsion, fully residually free groups, fundamental groups of hyperbolic 3-manifolds.*

The existence of a polynomial time algorithm for the compressed word problem of a fully residually free groups was also shown by Macdonald [120].

5.2 Trace Monoids

5.2.1 General Definitions

Our approach to the compressed word problem for graph products will be based on the theory of traces (partially commutative words). In the following we introduce some notions from trace theory; see [49, 52] for more details. Let us fix an independence alphabet (Σ, I) and let (Σ, D) be the corresponding dependence alphabet; see Definition 5.1. The set Σ may be infinite, but most of the time, it will be finite in this chapter.

Definition 5.8 (trace monoid). The *trace monoid* $\mathbb{M}(\Sigma, I)$ is defined as the quotient monoid $\mathbb{M}(\Sigma, I) = \Sigma^* / \{(ab, ba) \mid (a, b) \in I\}$. Its elements are called *traces*.

Trace monoids are known to be cancellative. We denote by $[w]_I$ the trace represented by the word $w \in \Sigma^*$. The trace $[\varepsilon]_I$ is the empty trace; it is the identity of the monoid $\mathbb{M}(\Sigma, I)$ and we denote it simply by ε. Since the relations (ab, ba) do not change the length or the alphabet of a word, we can define $\mathsf{alph}([w]_I) = \mathsf{alph}(w)$ and $|[w]_I| = |w|$. For $a \in \Sigma$ let $I(a) = \{b \in \Sigma \mid (a, b) \in I\}$ be the letters that commute with a and $D(a) = \Sigma \setminus I(a)$. For traces $u, v \in \mathbb{M}(\Sigma, I)$ with $\mathsf{alph}(u) \times \mathsf{alph}(v) \subseteq I$ we also write $u I v$.

Definition 5.9 (independence clique). An *independence clique* is a subset $\Delta \subseteq \Sigma$ such that $(a, b) \in I$ for all $a, b \in \Delta$ with $a \neq b$. For a *finite* independence clique Δ, we write $[\Delta]_I$ for the trace $[a_1 a_2 \cdots a_n]_I$, where $a_1, a_2, \ldots, a_n$ is an arbitrary enumeration of Δ (the precise enumeration is not important).

The following lemma is known as Levi's Lemma. It is one of the most fundamental facts for trace monoids; see, e.g., [52, p 74].

Lemma 5.10 (Levi's Lemma). *Let $u_1, \ldots, u_m, v_1, \ldots, v_n \in \mathbb{M}$. Then $u_1 u_2 \cdots u_m = v_1 v_2 \cdots v_n$ if and only if there exist $w_{i,j} \in \mathbb{M}$ ($1 \leq i \leq m, 1 \leq j \leq n$) such that*

- $u_i = w_{i,1} w_{i,2} \cdots w_{i,n}$ *for every* $1 \leq i \leq m$,
- $v_j = w_{1,j} w_{2,j} \cdots w_{m,j}$ *for every* $1 \leq j \leq n$, *and*
- $(w_{i,j}, w_{k,\ell}) \in I$ *if* $1 \leq i < k \leq m$ *and* $n \geq j > \ell \geq 1$.

The situation in the lemma will be visualized by a diagram of the following kind. The ith column corresponds to u_i, the jth row corresponds to v_j, and the intersection of the ith column and the jth row represents $w_{i,j}$. Furthermore $w_{i,j}$ and $w_{k,\ell}$ are independent if one of them is left above the other one.

v_n	$w_{1,n}$	$w_{2,n}$	$w_{3,n}$	$\cdots$	$w_{m,n}$
$\vdots$	$\vdots$	$\vdots$	$\vdots$	$\vdots$	$\vdots$
v_3	$w_{1,3}$	$w_{2,3}$	$w_{3,3}$	$\cdots$	$w_{m,3}$
v_2	$w_{1,2}$	$w_{2,2}$	$w_{3,2}$	$\cdots$	$w_{m,2}$
v_1	$w_{1,1}$	$w_{2,1}$	$w_{3,1}$	$\cdots$	$w_{m,1}$
	u_1	u_2	u_3	$\cdots$	u_m

A convenient representation for traces are *dependence graphs*, which are node-labeled directed acyclic graphs. For a word $w \in \Sigma^*$ the dependence graph D_w has vertex set $\{1, \ldots, |w|\}$ where the node i is labeled with $w[i]$. There is an edge from vertex i to j if and only if $i < j$ and $(w[i], w[j]) \in D$. It is easy to see that for two words $w, w' \in \Sigma^*$ we have $[w]_I = [w']_I$ if and only if D_w and $D_{w'}$ are isomorphic node-labeled graphs. Hence, we can speak of *the* dependence graph of a trace.

Example 5.11. We consider the following independence alphabet (Σ, I):

c ———— a

e ——— d ——— b

Then the corresponding dependence alphabet is

a   e

b ——— c ——— d

We consider the words $u = aeadbacdd$ and $v = eaabdcaeb$. Then the dependence graphs D_u of u and D_v of v look as follows, where we label the vertex i with the letter $u[i]$ (respectively, $v[i]$):

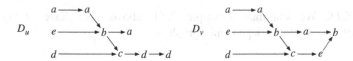

Note that we only show Hasse diagrams and hence omit, for instance, the edge from the first d to the last d in D_u.

Definition 5.12 (downward-closed set, convex set). Let E_w be the edge relation for the dependence graph D_w for a trace w. A subset $V \subseteq \{1, \dots, |w|\}$ is called *downward closed* if $(i, j) \in E_w$ and $j \in V$ implies $i \in V$. A subset $V \subseteq \{1, \dots, |w|\}$ is called *convex* if $(i, j), (j, k) \in E_w^*$ and $i, k \in V$ implies $j \in V$.

Definition 5.13 (projection homomorphism). For $\Delta \subseteq \Sigma$ we define the *projection homomorphism* $\pi_\Delta : \mathbb{M}(\Sigma, I) \to \mathbb{M}(\Delta, I \cap (\Delta \times \Delta))$ by $\pi_\Delta(a) = \varepsilon$ for $a \in \Sigma \setminus \Delta$ and $\pi_\Delta(a) = a$ for $a \in \Delta$.

With this definition, we get the following projection lemma.

Lemma 5.14. *For $u, v \in \mathbb{M}(\Sigma, I)$ we have $u = v$ if and only if $\pi_{\{a,b\}}(u) = \pi_{\{a,b\}}(v)$ for all $(a, b) \in D$.*

5.2.2 The Prefix and Suffix Order on Traces

Definition 5.15 (prefix and suffix of a trace). Let $u, v \in \mathbb{M}(\Sigma, I)$. Then u is a *prefix* (respectively, *suffix*) of v if there exists some $w \in \mathbb{M}(\Sigma, I)$ such that $uw = v$ (respectively, $wu = v$) in $\mathbb{M}(\Sigma, I)$, for short $u \preceq_p v$ (respectively, $u \preceq_s v$).

Prefixes of a trace u exactly correspond to downward-closed subsets of the dependence graph of u.

Definition 5.16 ($u \sqcap_p v$, $u \sqcap_s v$, $u \setminus_p v$, $u \setminus_s v$)**.** The *prefix infimum* (respectively, *suffix infimum*) $u \sqcap_p v$ (respectively, $u \sqcap_s v$) is the largest trace w with respect to $\preceq_p$ (respectively, $\preceq_s$) such that $w \preceq_p u$ and $w \preceq_p v$ (respectively, $w \preceq_s u$ and $w \preceq_s v$); it always exists [41]. With $u \setminus_p v$ (respectively, $u \setminus_s v$) we denote the unique trace w such that $u = (u \sqcap_p v)w$ (respectively, $u = w(u \sqcap_s v)$). Uniqueness follows from the fact that $\mathbb{M}(\Sigma, I)$ is cancellative.

Note that $u \setminus_p v = u \setminus_p (u \sqcap_p v)$ and $u \setminus_s v = u \setminus_s (u \sqcap_s v)$.

Definition 5.17 ($\min(u)$ and $\max(u)$)**.** For $u \in \mathbb{M}(\Sigma, I)$, we define

$$\min(u) = \{a \in \Sigma \mid a \preceq_p u\} \text{ and}$$

$$\max(u) = \{a \in \Sigma \mid a \preceq_s u\}.$$

Clearly, $\min(u)$ and $\max(u)$ are finite independence cliques and $[\min(u)]_I \preceq_p u$ and $[\max(u)]_I \preceq_s u$. Occasionally, we will identify the traces $[\min(u)]_I$ and $[\max(u)]_I$ with the independence cliques $\min(u)$ and $\max(u)$, respectively.

Example 5.18. We continue Example 5.11 above. We have $u \sqcap_p v = [aeadbac]_I =: w$ and its dependence graph is

Furthermore we have $\min(w) = \{a, d, e\}$ and $\max(w) = \{a, c\}$.

We also need the following lemma from [109].

Lemma 5.19. *For $u, v \in \mathbb{M}(\Sigma, I)$ we have $u \preceq_p v$ if and only if the word $\pi_{\{a,b\}}(u)$ is a prefix of the word $\pi_{\{a,b\}}(v)$ for all $(a, b) \in D$.*

5.2.3 Trace Rewriting Systems

Trace rewriting systems are defined analogously to semi-Thue systems (i.e., word rewriting systems).

Definition 5.20 (trace rewriting system). A *trace rewriting system* R over the trace monoid $\mathbb{M}(\Sigma, I)$ is a finite subset of $\mathbb{M}(\Sigma, I) \times \mathbb{M}(\Sigma, I)$ [49]. We define the *one-step rewrite relation* $\to_R \subseteq \mathbb{M}(\Sigma, I) \times \mathbb{M}(\Sigma, I)$ by $x \to_R y$ if and only if there are $u, v \in \mathbb{M}(\Sigma, I)$ and $(\ell, r) \in R$ such that $x = u\ell v$ and $y = urv$.

As for semi-Thue systems, we say that the trace rewriting system R is Noetherian (confluent, locally confluent) if $\to_R$ is Noetherian (confluent, locally confluent). Let $\mathsf{IRR}(R) = \mathsf{IRR}(\to_R)$ (the set of *irreducible traces* with respect to R) and

$\mathrm{NF}_R(w) = \mathrm{NF}_{\to_R}(w)$ (if R is Noetherian and confluent) be the *normal form* of the trace w. Trace rewriting systems are studied in detail in [49].

5.2.4 Confluent and Terminating Trace Rewriting Systems for Graph Products

Let us fix for this subsection a *finite* independence alphabet (W, E) with $W = \{1, \ldots, n\}$ and finitely generated groups G_i for $i \in \{1, \ldots, n\}$. For pairwise disjoint nonempty sets $C_1, \ldots, C_n$ we define the independence relation

$$E[C_1, \ldots, C_n] = \bigcup_{(i,j) \in E} C_i \times C_j \tag{5.1}$$

on the alphabet $\bigcup_{i=1}^n C_i$. Every independence clique of $(\bigcup_{i=1}^n C_i, E[C_1, \ldots, C_n])$ has size at most n. We define a (possibly infinite) independence alphabet as in [51, 100]: let

$$A_i = G_i \setminus \{1\} \quad \text{and} \quad A = \bigcup_{i=1}^n A_i.$$

We assume that $A_1, \ldots, A_n$ are pairwise disjoint. We fix the independence relation

$$I = E[A_1, \ldots, A_n]$$

on A for the rest of this subsection. The independence alphabet (A, I) is the only independence alphabet in this chapter, which may be infinite. On $\mathbb{M}(A, I)$, we define the trace rewriting system

$$R = \bigcup_{i=1}^n \Big(\{([aa^{-1}]_I, \varepsilon) \mid a \in A_i\} \cup \{([ab]_I, [c]_I) \mid a, b, c \in A_i, ab = c \text{ in } G_i\} \Big). \tag{5.2}$$

Clearly, R is terminating (it reduces the length of traces). The following lemma was shown in [100].

Lemma 5.21. *The trace rewriting system R is confluent.*

Proof. Since R is terminating, it suffices by Newman's Lemma [134] to show that R is locally confluent, i.e., for all $s, s_1, s_2 \in \mathbb{M}(A, I)$ with $s \to_R s_1$ and $s \to_R s_2$ there exists $s' \in \mathbb{M}(A, I)$ with $s_1 \to_R^* s'$ and $s_2 \to_R^* s'$.

Thus, assume that $s \to_R s_1$ and $s \to_R s_2$. Hence, $s = t_i a_i b_i u_i$ and $s_i = t_i r_i u_i$ for $i \in \{1, 2\}$, where $(a_i b_i, r_i) \in R$. Thus, $r_i \in A \cup \{\varepsilon\}$. By applying Levi's Lemma 5.10 to the identity $t_1 a_1 b_1 u_1 = t_2 a_2 b_2 u_2$, we obtain the following diagram:

u_2	w_2	q_1	v_2
a_2b_2	p_2	t	q_2
t_2	v_1	p_1	w_1
	t_1	a_1b_1	u_1

Thus, $w_1 I w_2$. Since the trace monoid is cancellative, we can assume $v_1 = v_2 = \varepsilon$ for the further arguments. Assume that $a_i, b_i \in A_{\sigma_i}$ with $\sigma_i \in \{1, \ldots, n\}$. Let us first consider the case $t \neq \varepsilon$. Thus, $\sigma_1 = \sigma_2 = \sigma$, $r_1, r_2 \in A_\sigma \cup \{\varepsilon\}$, and $c I w_i$ for all $c \in A_\sigma$ and $i \in \{1, 2\}$. Moreover, since $p_1 I p_2$ but both traces only contain symbols from A_σ, we have $p_1 = \varepsilon$ or $p_2 = \varepsilon$ and similarly $q_1 = \varepsilon$ or $q_2 = \varepsilon$. If $p_1 = p_2 = q_1 = q_2 = \varepsilon$, then $a_1b_1 = t = a_2b_2$ and hence $r_1 = r_2$. We get $s_1 = w_2r_1w_1 = w_1r_2w_2 = s_2$. Otherwise, since a_1b_1 cannot be a proper factor of a_2b_2 and vice versa, we obtain up to symmetry the following diagram (recall that we assume $v_1 = v_2 = \varepsilon$):

u_2	w_2	b_1	ε
a_2b_2	a_2	$b_2 = a_1$	ε
t_2	ε	ε	w_1
	t_1	a_1b_1	u_1

Thus, $s_1 = a_2w_2r_1w_1 = w_1a_2r_1w_2$ and $s_2 = w_1r_2w_2b_1 = w_1r_2b_1w_2$. In the group G_σ we have $a_2r_1 = a_2a_1b_1 = a_2b_2b_1 = r_2b_1$. But this implies that s_1 and s_2 can be reduced to the same trace. This concludes the case $t \neq \varepsilon$.

Now assume that $t = \varepsilon$. Thus, we have the following diagram:

u_2	w_2	q_1	ε
a_2b_2	p_2	ε	q_2
t_2	ε	p_1	w_1
	t_1	a_1b_1	u_1

If also $p_1 = \varepsilon$, the diagram looks as follows:

u_2	w_2	a_1b_1	ε
a_2b_2	p_2	ε	q_2
t_2	ε	ε	w_1
	t_1	a_1b_1	u_1

In particular, $w_1q_2 I a_1$ implying $w_1q_2 I r_1$. We have to show that $s_1 = p_2w_2r_1w_1q_2$ and $s_2 = w_1r_2w_2a_1b_1$ can be reduced to the same trace. We have $s_2 \to_R w_1r_2w_2r_1$. Moreover with the independencies listed above, we obtain

$$s_1 = p_2w_2r_1w_1q_2 = p_2w_2w_1q_2r_1 = w_1p_2q_2w_2r_1 \to_R w_1r_2w_2r_1.$$

If one of the traces p_2, q_1, or q_2 is empty, then we can argue analogously. Thus, we may assume that p_1, p_2, q_1, and q_2 are nonempty. It follows $p_1 = a_1$, $q_1 = b_1$, $p_2 = a_2$, and $q_2 = b_2$. Then all traces from $\{w_1, w_2, a_1b_1, a_2b_2\}$ are pairwise independent, from which it follows again easily that s_1 and s_2 can be reduced to $w_1 w_2 r_1 r_2$. $\qquad\square$

Since R is terminating and confluent it defines unique normal forms. Alternatively to Definition 5.2, one can define the graph product $\mathbb{G}(W, E, (G_i)_{i \in W})$ of $G_1, \ldots, G_n$ as the quotient monoid

$$\mathbb{G}(W, E, (G_i)_{i \in W}) = \mathbb{M}(A, I)/R.$$

The following lemma is important for solving the word problem in a graph product $G = \mathbb{G}(W, E, (G_i)_{i \in W})$.

Lemma 5.22. *Let $u, v \in A^*$. Then $u = v$ in G if and only if $\mathsf{NF}_R([u]_I) = \mathsf{NF}_R([v]_I)$. In particular we have $u = 1$ in G if and only if $\mathsf{NF}_R([u]_I) = \varepsilon$.*

Proof. The if-direction is trivial. Let on the other hand $u, v \in A^*$ and suppose that $u = v$ in G. By definition this is the case if and only if $[u]_I$ and $[v]_I$ represent the same element from $\mathbb{M}(A, I)/R$ and are hence congruent. Since R produces a normal form for elements from the same congruence class, this implies that $\mathsf{NF}_R([u]_I) = \mathsf{NF}_R([v]_I)$. $\qquad\square$

On the trace monoid $\mathbb{M}(A, I)$ we can define a natural involution $^{-1} : \mathbb{M}(A, I) \to \mathbb{M}(A, I)$. For a letter $a \in A_\sigma$, a^{-1} is the inverse of a in the group G_σ. This defines an involution on A, which can be extended to the trace monoid $\mathbb{M}(A, I)$ by $[a_1 \cdots a_n]_I^{-1} = [a_n^{-1} \cdots a_1^{-1}]_I$.

For the normal form of the product of two R-irreducible traces we have the following lemma, which was shown in [51] (equation (21) in the proof of Lemma 22) using a slightly different notation.

Lemma 5.23. *Let $u, v \in \mathsf{IRR}(R)$ be irreducible traces from $\mathbb{M}(A, I)$. Let $x = u \setminus_s v^{-1}$, $y = v \setminus_p u^{-1}$, $x' = x \setminus_s \max(x)$, and $y' = y \setminus_p \min(y)$. Then*

$$\mathsf{NF}_R(uv) = x'\mathsf{NF}_R(\max(x)\min(y))y'.$$

Before we prove this lemma, let us give some intuition. Since u and v both belong to $\mathsf{IRR}(R)$, rules of R can be only applied at the border of u and v. We have $u = x(u \sqcap_s v^{-1})$ and $v = (u^{-1} \sqcap_p v)y$. Moreover, $(u \sqcap_s v^{-1})^{-1} = (u^{-1} \sqcap_p v)$ and $(u^{-1} \sqcap_p v)$ is the longest trace p for which there exist traces x and y with $u = xp^{-1}$ and $v = py$. This means that $uv \to_R^* xy$, and in the trace xy no rule of the form $([aa^{-1}]_I, \varepsilon)$ can be applied. Hence, only rules of the form $([ab]_I, c) \in R$ can be applied and rule application can only occur at the border between x and y.

Proof of Lemma 5.23. As remarked above, we have $u = x(u \sqcap_s v^{-1})$ and $v = (u^{-1} \sqcap_p v)y$. Moreover, $(u \sqcap_s v^{-1})^{-1} = (u^{-1} \sqcap_p v)$ and $(u^{-1} \sqcap_p v)$ is the longest trace p for which there exist traces x and y with $u = xp^{-1}$ and $v = py$.

With x' and y' as defined in the lemma we get

$$uv = x' \max(x)(u \sqcap_s v^{-1})(u^{-1} \sqcap_p v) \max(y)y' \to_R^* x' \max(x) \max(y)y'$$

$$\to_R^* x' \mathsf{NF}_R(\max(x) \min(y))y'.$$

Let $q = \mathsf{NF}_R(\max(x) \min(y)) \in \mathsf{IRR}(R)$. It remains to show that $x'qy' \in \mathsf{IRR}(R)$. Note that also $x', y' \in \mathsf{IRR}(R)$ (since x' is a prefix of $u \in \mathsf{IRR}(R)$ and y' is a suffix of $v \in \mathsf{IRR}(R)$).

Recall that $\max(x)$ and $\min(y)$ are independence cliques. Moreover, there cannot exist $a \in \max(x)$ with $a^{-1} \in \min(y)$, because this a could be moved to $(u^{-1} \sqcap_p v)$. It follows that q contains a symbol from A_σ if and only if $A_\sigma \cap (\max(x) \cup \min(y)) \neq \emptyset$.

In order to get a contradiction, let us assume that there exist $([ab]_I, r) \in R$ and traces q_1, q_2 such that $x'qy' = q_1 abq_2$. By Levi's Lemma 5.10 and $x', q, y' \in \mathsf{IRR}(R)$ we obtain up to symmetry one of the following two diagrams:

z_2	s_2	q_2	t_2
ab	a	ε	b
z_1	s_1	q_1	t_1
	x'	q	y'

z_2	s_2	q_2	t_2
ab	a	b	ε
z_1	s_1	q_1	t_1
	x'	q	y'

Assume that $a, b \in A_\sigma$ ($\sigma \in W$). Let us first consider the left diagram. Since $a I q_1$, $b I q_2$, and $q = q_1 q_2$, we obtain $a I q$ and thus $a I \max(x)$. Furthermore, from the diagram we obtain also $b I s_2$. Thus, $a I s_2$, which implies $a \in \max(x')$. Together with $a I \max(x)$ it follows that $a \in \max(x' \max(x)) = \max(x)$, which contradicts $a I \max(x)$.

Now let us consider the right diagram. Again we have $a \in \max(x')$. Furthermore, $a I q_1$, i.e., $b I q_1$. Hence, $b \in \min(q) \cap A_\sigma$. Recall that $q = \mathsf{NF}_R(\max(x) \min(y))$. Hence, there are two possibilities: Either there exists $a' \in \max(x) \cap A_\sigma$ or $b \in \min(y)$ and $b I \max(x)$. If $a' \in \max(x) \cap A_\sigma$, then $x = x' \max(x)$ would contain the factor $[aa']_I$, which contradicts $x \in \mathsf{IRR}(R)$. If $b \in \min(y)$ and $b I \max(x)$, then also $a I \max(x)$, which implies $a \in \max(x' \max(x)) = \max(x)$, the same contradiction as in the previous paragraph. $\square$

Note that in Lemma 5.23 $|\max(x)|$ as well as $|\min(y)|$ are bounded by $n = |W|$. Hence, there are at most n rewrite steps in the derivation of $\mathsf{NF}_R(\max(x) \min(y))$ from $\max(x) \min(y)$.

5.2.5 Simple Facts for Compressed Traces

SLPs allow to represent traces in a succinct way. In this section, we collect some tools for manipulating SLP-represented traces. These tools will be important in order to solve the compressed word problem for a graph product.

Lemma 5.24. *The following problem can be decided in polynomial time:*

input: A finite independence alphabet (Σ, I) and SLPs $\mathbb{A}$ and $\mathbb{B}$ over the terminal alphabet Σ.

question: Does $[\mathsf{val}(\mathbb{A})]_I = [\mathsf{val}(\mathbb{B})]_I$ hold?

Proof. From $\mathbb{A}$ and $\mathbb{B}$ we can compute by Proposition 3.9(4) in polynomial time for all $(a, b) \in D$ SLPs $\mathbb{A}_{a,b}$ and $\mathbb{B}_{a,b}$ with $\mathsf{val}(\mathbb{A}_{a,b}) = \pi_{\{a,b\}}(\mathsf{val}(\mathbb{A}))$ and $\mathsf{val}(\mathbb{B}_{a,b}) = \pi_{\{a,b\}}(\mathsf{val}(\mathbb{B}))$. By Lemma 5.14, it suffices to check $\mathsf{val}(\mathbb{A}_{a,b}) = \mathsf{val}(\mathbb{B}_{a,b})$, which is possible in polynomial time by Theorem 3.17. □

The previous lemma is slightly generalized by the following one.

Lemma 5.25. *The following problem can be decided in polynomial time:*

input: A finite independence alphabet (Σ, I) and SLPs $\mathbb{A}$ and $\mathbb{B}$ over the terminal alphabet Σ.

question: Does $[\mathsf{val}(\mathbb{A})]_I \preceq_p [\mathsf{val}(\mathbb{B})]_I$ hold?

Proof. Compute $\mathbb{A}_{a,b}$ and $\mathbb{B}_{a,b}$ as above. By Lemma 5.19, we have to check for all $(a, b) \in D$ whether the word $\mathsf{val}(\mathbb{A}_{a,b})$ is a prefix of the word $\mathsf{val}(\mathbb{B}_{a,b})$. But this can be easily reduced to an equivalence check: Compute in polynomial time $\ell_{a,b} = |\mathsf{val}(\mathbb{A}_{a,b})|$ (using Proposition 3.9(2)) and an SLP $\mathbb{C}_{a,b}$ with $\mathsf{val}(\mathbb{C}_{a,b}) = \mathsf{val}(\mathbb{B}_{a,b})[: \ell_{a,b}]$ (using Theorem 3.14). Finally check in polynomial time whether $\mathsf{val}(\mathbb{C}_{a,b}) = \mathsf{val}(\mathbb{A}_{a,b})$ for all $(a, b) \in D$ using Theorem 3.17. □

Lemma 5.26. *There is a polynomial time algorithm for the following problem:*

input: A finite independence alphabet (Σ, I) and an SLP $\mathbb{A}$ over the alphabet Σ

output: The sets $\max([\mathsf{val}(\mathbb{A})]_I)$ and $\min([\mathsf{val}(\mathbb{A})]_I)$

Proof. Without loss of generality we can assume that $\mathbb{A} = (V, \Sigma, S, \mathrm{rhs})$ is in Chomsky normal form. We show how to compute $\max([\mathsf{val}(\mathbb{A})]_I)$. First we compute $\mathsf{alph}(\mathsf{val}(\mathbb{A}))$ in polynomial time using Proposition 3.9(1). For $a \in \mathsf{alph}(\mathsf{val}(\mathbb{A}))$ let $p_a \in \{1, \ldots, |\mathsf{val}(\mathbb{A})|\}$ maximal such that $\mathsf{val}(\mathbb{A})[p_a] = a$. These numbers can be computed in polynomial time by Proposition 3.9(5). Using Theorem 3.14 we compute in polynomial time an SLP $\mathbb{B}$ such that $\mathsf{val}(\mathbb{B}) = \mathsf{val}(\mathbb{A})[p_a + 1 :]$ Then $a \in \max([\mathsf{val}(\mathbb{A})]_I)$ if and only if $a \; I \; \mathsf{alph}(\mathsf{val}(\mathbb{B}))$. This property can be checked in polynomial time by computing $\mathsf{alph}(\mathsf{val}(\mathbb{B}))$ (using Proposition 3.9(1)). Repeating this procedure for all $a \in \mathsf{alph}(\mathsf{val}(\mathbb{A}))$ we get the set $\max([\mathsf{val}(\mathbb{A})]_I)$. The set $\min([\mathsf{val}(\mathbb{A})]_I)$ can be determined similarly. □

For the following, recall the definition of a C-expression (Definition 3.3).

Lemma 5.27. *There is a polynomial time algorithm for the following problem:*

input: A finite independence alphabet (Σ, I) and an SLP $\mathbb{A}$ over the terminal alphabet Σ

output: C-expressions α and β with $\mathsf{Var}(\alpha) = \mathsf{Var}(\beta) = \{S\}$ (where S is the start variable of $\mathbb{A}$), $[\mathsf{val}_{\mathbb{A}}(\alpha)]_I = [\mathsf{val}(\mathbb{A})]_I \setminus_s \max([\mathsf{val}(\mathbb{A})]_I)$, and $[\mathsf{val}_{\mathbb{A}}(\beta)]_I = [\mathsf{val}(\mathbb{A})]_I \setminus_p \min([\mathsf{val}(\mathbb{A})]_I)$

Moreover, $|\alpha|$ (respectively, $|\beta|$) can be bounded by $\mathcal{O}(|\min([\mathsf{val}(\mathbb{A})]_I)| \cdot \log_2(|\mathsf{val}(\mathbb{A})|))$ (respectively, $\mathcal{O}(|\max([\mathsf{val}(\mathbb{A})]_I)| \cdot \log_2(|\mathsf{val}(\mathbb{A})|))$).

Proof. We show how to compute the expression α in polynomial time; for β one can argue analogously. By Lemma 5.26 we can find the set $\max([\mathsf{val}(\mathbb{A})]_I)$ in polynomial time. Let $p_a \in \{1, \ldots, |\mathsf{val}(\mathbb{A})|\}$ be maximal such that $\mathsf{val}(\mathbb{A})[p_a] = a$ for $a \in \max([\mathsf{val}(\mathbb{A})]_I)$ and let $\{k_1, \ldots, k_m\} = \{p_a \mid a \in \max([\mathsf{val}(\mathbb{A})]_I)\}$ with $k_1 < k_2 < \cdots < k_m$. These numbers can be computed in polynomial time by Proposition 3.9(5). We set

$$\alpha = S[: k_1 - 1]S[k_1 + 1 : k_2 - 1] \cdots S[k_{m-1} + 1 : k_m - 1]S[k_m + 1 :].$$

Then $[\mathsf{val}_{\mathbb{A}}(\alpha)]_I = [\mathsf{val}(\mathbb{A})]_I \setminus_s \max([\mathsf{val}(\mathbb{A})]_I)$. Since the positions $k_1, \ldots, k_m$ are represented in binary, each of them needs $\mathcal{O}(\log_2(|\mathsf{val}(\mathbb{A})|))$ many bits. Hence $|\alpha|$ can be bounded by $\mathcal{O}(m \cdot \log_2(|\mathsf{val}(\mathbb{A})|))$. Since $m = |\max([\mathsf{val}(\mathbb{A})]_I)|$ we have $|\alpha| \leq \mathcal{O}(|\max([\mathsf{val}(\mathbb{A})]_I)| \cdot \log_2(|\mathsf{val}(\mathbb{A})|))$. $\square$

5.3   The Compressed Word Problem for Graph Products

In this section, we will prove Theorem 5.4. Our technique will generalize our algorithm for free groups (Theorem 4.11). There are two aspects which make the proof of Theorem 5.4 more complicated than the proof of Theorem 4.11 for free groups:

- We have to deal with partial commutation, i.e., traces instead of words.
- We have to prove a preservation theorem: Whereas Theorem 4.11 deals with a fixed group (a finitely generated free group), the G_i ($i \in W$) in Theorem 5.4 are arbitrary finitely generated groups and we have to reduce the compressed word problem for the graph product to the compressed word problems for the groups G_i.

Let us fix the *finite* independence alphabet (W, E) with $W = \{1, \ldots, n\}$ and finitely generated groups G_i for $i \in \{1, \ldots, n\}$ for the rest of this section. Furthermore, let Σ_i be a finite generating set for G_i for $i \in \{1, \ldots, n\}$. Without loss of generality we can assume that Σ_i does not contain the identity element and that $\Sigma_i \cap \Sigma_j = \emptyset$ for $i \neq j$. We define $\Sigma = \bigcup_{i=1}^n \Sigma_i$. Let G denote the graph product $\mathbb{G}(W, E, (G_i)_{i \in W})$ for the rest of this section. Moreover, let A_i, A, I, and R have the same meaning as in Sect. 5.2.4. Note that $\Sigma_i \subseteq A_i$ for all $1 \leq i \leq n$.

Recall the definition of a 2-level PCSLP (Definition 3.24). For the following discussion, let us fix a 2-level PCSLP

$$\mathbb{B} = (\mathsf{Up}, \mathsf{Lo}, \Sigma \cup \Sigma^{-1}, S, \mathsf{rhs})$$

over the terminal alphabet $\Sigma \cup \Sigma^{-1}$ (the monoid generating set of our graph product G). We introduce several properties for $\mathbb{B}$.

Definition 5.28 (pure). The 2-level PCSLP $\mathbb{B}$ is *pure* if for every $X \in \mathsf{Lo}$ there exists $i \in W$ such that $\mathsf{val}_{\mathbb{B}}(X) \in (\Sigma_i \cup \Sigma_i^{-1})^+$ and $\mathsf{val}_{\mathbb{B}}(X) \neq 1$ in G_i (hence, $\mathsf{val}_{\mathbb{B}}(X)$ represents a group element from the set A).

For the following notations, assume that $\mathbb{B}$ is pure. Then, we can define the mapping $\mathsf{type}_{\mathbb{B}} : \mathsf{Lo} \to W$ by $\mathsf{type}_{\mathbb{B}}(X) = i$ if $\mathsf{val}_{\mathbb{B}}(X) \in (\Sigma_i \cup \Sigma_i^{-1})^+$. For $i \in W$ let

$$\mathsf{Lo}(i) = \{X \in \mathsf{Lo} \mid \mathsf{type}_{\mathbb{B}}(X) = i\}.$$

Then the sets $\mathsf{Lo}(1), \ldots, \mathsf{Lo}(n)$ form a partition of Lo. Moreover, using (5.1) on page 93 we can define an independence relation $I_{\mathbb{B}}$ on Lo by

$$I_{\mathbb{B}} = E[\mathsf{Lo}(1), \ldots, \mathsf{Lo}(n)].$$

Definition 5.29 (nicely projecting). The 2-level PCSLP $\mathbb{B}$ is *nicely projecting* if for every subexpression of the form $\pi_\Delta(\alpha)$ ($\Delta \subseteq \mathsf{Lo}$) that appears in a right-hand side of $\mathsf{up}(\mathbb{B})$, there exists $K \subseteq W$ with $\Delta = \bigcup_{i \in K} \mathsf{Lo}(i)$.

This condition will be needed in order to apply Lemma 3.16. Note that the number of all sets $\bigcup_{i \in K} \mathsf{Lo}(i)$ with $K \subseteq W$ is bounded by $2^n = \mathcal{O}(1)$ (the size of W is a fixed constant in our consideration).

Definition 5.30 (irredundant). The 2-level PCSLP $\mathbb{B}$ is *irredundant* if for all $X, Y \in \mathsf{Lo}$ such that $X \neq Y$ and $\mathsf{type}_{\mathbb{B}}(X) = \mathsf{type}_{\mathbb{B}}(Y) = i$, we have $\mathsf{val}_{\mathbb{B}}(X) \neq \mathsf{val}_{\mathbb{B}}(Y)$ in the group G_i.

One can think of a pure and irredundant 2-level PCSLP $\mathbb{B}$ as a PCSLP, where the terminal alphabet is a finite subset $B \subseteq A$, with $A = \bigcup_{i \in W} G_i \setminus \{1\}$ from Sect. 5.2.4: Take $\mathsf{up}(\mathbb{B})$, replace every variable $X \in \mathsf{Lo}(i)$ ($i \in W$) by the group element from G_i represented by $\mathsf{val}_{\mathbb{B}}(X)$, and remove the lower part $\mathsf{lo}(\mathbb{B})$. Moreover, each element from $B \cap A_i$ ($i \in W$) is represented by a unique SLP over the terminal alphabet $\Sigma_i \cup \Sigma_i^{-1}$ (namely, the lower part $\mathsf{lo}(\mathbb{B})$ with the appropriate start variable). If $\mathbb{B}$ is pure but not irredundant, then using the compressed word problems for the groups G_i as oracles, one can compute a pure and irredundant 2-level PCSLP $\mathbb{C}$ such that $\mathsf{val}(\mathbb{B}) = \mathsf{val}(\mathbb{C})$ in G as follows: If $\mathbb{B}$ contains two variables $X, Y \in \mathsf{Lo}$ such that $X \neq Y$, $\mathsf{type}_{\mathbb{B}}(X) = \mathsf{type}_{\mathbb{B}}(Y) = i$ and $\mathsf{val}_{\mathbb{B}}(X) = \mathsf{val}_{\mathbb{B}}(Y)$ in G_i, one has to replace Y in all right-hand sides by X. Note that this process does not change the set of upper level variables of $\mathbb{B}$.

Definition 5.31 (saturated). The 2-level PCSLP $\mathbb{B}$ is *saturated* if for every $X \in \mathsf{Lo}$ with $\mathsf{type}_{\mathbb{B}}(X) = i$, there exists $Y \in \mathsf{Lo}$ with $\mathsf{type}_{\mathbb{B}}(Y) = i$ and $\mathsf{val}_{\mathbb{B}}(Y) = \mathsf{val}_{\mathbb{B}}(X)^{-1}$ in G_i.

If $\mathbb{B}$ is pure, irredundant, and saturated, then for every $X \in \mathsf{Lo}$ with $\mathsf{type}_{\mathbb{B}}(X) = i$, there must be a unique $Y \in \mathsf{Lo}$ with $\mathsf{type}_{\mathbb{B}}(Y) = i$ and $\mathsf{val}_{\mathbb{B}}(Y) = \mathsf{val}_{\mathbb{B}}(X)^{-1}$ in G_i (we may have $Y = X$ in case $\mathsf{val}_{\mathbb{B}}(X)^2 = 1$ in G_i). This Y is denoted by X^{-1}, and we define $(X_1 \cdots X_n)^{-1} = X_n^{-1} \cdots X_1^{-1}$ for $X_1, \ldots, X_n \in \mathsf{Lo}$.

Definition 5.32 (well-formed). The 2-level PCSLP $\mathbb{B}$ is *well formed* if it is pure, irredundant, saturated, and nicely projecting.

Assume that $\mathbb{B}$ is well formed. We call a trace $w \in \mathbb{M}(\mathsf{Lo}, I_\mathbb{B})$ *reduced* if it contains no factor $[YZ]_{I_\mathbb{B}}$ with $Y, Z \in \mathsf{Lo}$ and $\mathsf{type}_\mathbb{B}(Y) = \mathsf{type}_\mathbb{B}(Z)$. Note that $[X_1 \cdots X_m]_{I_\mathbb{B}} \in \mathbb{M}(\mathsf{Lo}, I_\mathbb{B})$ with $X_1, \ldots, X_m \in \mathsf{Lo}$ is reduced if and only if $[a_1 \cdots a_m]_I \in \mathsf{IRR}(R)$, where $a_j \in A$ is the group element represented by $\mathsf{val}_\mathbb{B}(X_j)$ for $1 \leq j \leq m$. A variable $X \in \mathsf{Up} \cup \mathsf{Lo}$ is reduced if either $X \in \mathsf{Lo}$ or $X \in \mathsf{Up}$ and the trace $[\mathsf{uval}(X)]_{I_\mathbb{B}}$ is reduced. Finally, $\mathbb{B}$ is reduced if every variable X of $\mathbb{B}$ is reduced. We have

Lemma 5.33. *Let $\mathbb{B}$ be a well-formed and reduced 2-level PCSLP. Then $\mathsf{val}(\mathbb{B}) = 1$ in G if and only if $\mathsf{uval}(\mathbb{B}) = \varepsilon$.*

Proof. Clearly, if $\mathsf{uval}(\mathbb{B}) = \varepsilon$, then also $\mathsf{val}(\mathbb{B}) = \varepsilon$ and hence $\mathsf{val}(\mathbb{B}) = 1$ in G. For the other direction we assume that $\mathsf{uval}(\mathbb{B}) = X_1 \cdots X_m$ for some $m > 0$. Since $\mathbb{B}$ is pure there are $a_1, \ldots, a_m \in A$ such that $\mathsf{val}(X_i)$ represents the group element a_i for $i \in \{1, \ldots, m\}$. Since $\mathbb{B}$ is reduced, we have $[a_1 \cdots a_m]_I \in \mathsf{IRR}(R)$ and hence $\mathsf{NF}_R([a_1 \cdots a_m]_I) = [a_1 \cdots a_m]_I \neq \varepsilon$. From Lemma 5.22 it follows that $a_1 \cdots a_m \neq 1$ in G and hence $\mathsf{val}(\mathbb{B}) \neq 1$ in G. $\qquad\square$

Together with Lemma 5.33, the following proposition can be used to solve the compressed word problem for the graph product G.

Proposition 5.34. *Given an SLP $\mathbb{A}$ over $\Sigma \cup \Sigma^{-1}$ we can compute a well-formed and reduced 2-level PCSLP $\mathbb{B}$ with $\mathsf{val}(\mathbb{A}) = \mathsf{val}(\mathbb{B})$ in G in polynomial time using oracle access to the decision problems $\mathsf{CWP}(G_i)$ ($1 \leq i \leq n$).*

Proof. Let $\mathbb{A} = (V_\mathbb{A}, \Sigma \cup \Sigma^{-1}, S, \mathsf{rhs}_\mathbb{A})$ be the given input SLP over $\Sigma \cup \Sigma^{-1}$. We assume without loss of generality that $\mathbb{A}$ is in Chomsky normal form. Moreover, we exclude the trivial case that $\mathsf{rhs}_\mathbb{A}(S) \in \Sigma \cup \Sigma^{-1}$. We construct a sequence of 2-level PCSLPs $\mathbb{A}_j = (\mathsf{Up}_j, \mathsf{Lo}_j, \Sigma \cup \Sigma^{-1}, S, \mathsf{rhs}_j)$ ($0 \leq j \leq r \leq |V_\mathbb{A}|$) such that the following holds for all $0 \leq j \leq r$:

(a) $\mathbb{A}_j$ is well formed.
(b) $|\mathbb{A}_j| \leq 2 \cdot |\mathbb{A}| + \mathcal{O}(j \cdot |\mathbb{A}|) \leq 2 \cdot |\mathbb{A}| + \mathcal{O}(|V_\mathbb{A}| \cdot |\mathbb{A}|)$.
(c) $\mathsf{val}(\mathbb{A}) = \mathsf{val}(\mathbb{A}_j)$ in G for all $0 \leq j \leq r$.
(d) If $X \in \mathsf{Up}_j$ is not reduced, then $\mathsf{rhs}_j(X) \in (\mathsf{Up}_j \cup \mathsf{Lo}_j)^2$.
(f) $|\mathsf{uval}(\mathbb{A}_j)| \leq |\mathsf{val}(\mathbb{A})|$.

Moreover, the final 2-level PCSLP $\mathbb{B} = \mathbb{A}_r$ will be reduced. Let us write type_j for $\mathsf{type}_{\mathbb{A}_j}$ and I_j for $I_{\mathbb{A}_j}$ in the following.

During the construction of $\mathbb{A}_{j+1}$ from $\mathbb{A}_j$, we will replace the right-hand side YZ ($Y, Z \in \mathsf{Up}_j \cup \mathsf{Lo}_j$) for a non-reduced (with respect to $\mathbb{A}_j$) variable $X \in \mathsf{Up}_j$ by a new right-hand side of size $\mathcal{O}(|\mathbb{A}|)$, so that X is reduced in $\mathbb{A}_{j+1}$ and $\mathsf{val}_{\mathbb{A}_j}(X) = \mathsf{val}_{\mathbb{A}_{j+1}}(X)$ in G. All other right-hand sides for upper level variables will be kept, and constantly many new lower level variables will be added.

We start the construction with the 2-level PCSLP

$$(\{X \in V_\mathbb{A} \mid \text{rhs}_\mathbb{A}(X) \in V_\mathbb{A}^2\}, \{X \in V_\mathbb{A} \mid \text{rhs}_\mathbb{A}(X) \in \Sigma \cup \Sigma^{-1}\}, \Sigma \cup \Sigma^{-1}, S, \text{rhs}_\mathbb{A}).$$

Note that S is an upper level variable in this system (which is required for 2-level PCSLPs) since we assume $\text{rhs}_\mathbb{A}(S) \notin \Sigma \cup \Sigma^{-1}$. Moreover, the system is pure and nicely projecting (there are no projection operations in right-hand sides), but not necessarily irredundant and saturated. The latter two properties can be easily enforced by adding for every variable X with $\text{rhs}_\mathbb{A}(X) = a \in \Sigma \cup \Sigma^{-1}$, a variable X^{-1} for a^{-1} and then eliminating redundant lower level variables. The resulting 2-level PCSLP $\mathbb{A}_0$ is well formed and satisfies $|\mathbb{A}_0| \leq 2 \cdot |\mathbb{A}|$ and $\text{val}(\mathbb{A}_0) = \text{val}(\mathbb{A})$. Hence, (a), (b), and (c) are satisfied and also (d) and (e) clearly hold.

For the inductive step of the construction, assume that we have constructed $\mathbb{A}_j = (\text{Up}_j, \text{Lo}_j, \Sigma \cup \Sigma^{-1}, S, \text{rhs}_j)$ and let $X \in \text{Up}_j$, $Y, Z \in \text{Up}_j \cup \text{Lo}_j$ such that $\text{rhs}_j(X) = YZ$, X is not reduced, but Y and Z are already reduced. In order to make X reduced, we will apply Lemma 5.23. The following proposition, whose proof is postponed to the next Sect. 5.3.1, makes this application possible.

Proposition 5.35. *Let (W, E) be a fixed independence alphabet with $W = \{1, \ldots, n\}$. The following problem can be solved in polynomial time:*

input: Pairwise disjoint finite alphabets $\Gamma_1, \ldots, \Gamma_n$, an SLP $\mathbb{B}$ over the terminal alphabet $\Gamma = \bigcup_{i=1}^n \Gamma_i$, and two variables Y and Z from $\mathbb{B}$.
output: PC-expressions α, β with $\text{Var}(\alpha) = \text{Var}(\beta) = \{Y\}$ such that the following holds, where $J = E[\Gamma_1, \ldots, \Gamma_n]$:

(1) For every subexpression of the form $\pi_\Delta(\gamma)$ in α and β there exists $K \subseteq \{1, \ldots, n\}$ with $\Delta = \bigcup_{i \in K} \Gamma_i$.
(2) $[\text{val}_\mathbb{B}(\alpha)]_J = [\text{val}_\mathbb{B}(Y)]_J \setminus_p [\text{val}_\mathbb{B}(Z)]_J$
(3) $[\text{val}_\mathbb{B}(\beta)]_J = [\text{val}_\mathbb{B}(Y)]_J \sqcap_p [\text{val}_\mathbb{B}(Z)]_J$
(4) $|\alpha|, |\beta| \leq \mathcal{O}(\log |\text{val}(\mathbb{B})|).$

An analogous statement can be shown for the operations $\setminus_s$ and $\sqcap_s$ which refer to the suffix order on traces. Actually, we only need the PC-expression α for the trace difference.

In order to apply Proposition 5.35 to our situation we transform the upper level part $\text{up}(\mathbb{A}_j)$ into an equivalent SLP $\mathbb{C}$ over the terminal alphabet Lo_j using Lemma 3.16. This is possible, since $\mathbb{A}_j$ is nicely projecting by (a). Every upper level variable of $\mathbb{A}_j$ is also present in $\mathbb{C}$, and we have $\text{uval}_{\mathbb{A}_j}(C) = \text{val}_\mathbb{C}(C)$ for every $C \in \text{Up}_j$. The SLP $\mathbb{C}$ can be constructed in time polynomially bounded in $|\text{up}(\mathbb{A}_j)|$ and hence in $|\mathbb{A}|$. Note that we have $|\text{val}(\mathbb{C})| = |\text{uval}(\mathbb{A}_j)| \leq |\text{val}(\mathbb{A})| \leq 2^{|\mathbb{A}|}$ by (e). We can add to $\mathbb{C}$ symbols Y^{-1} and Z^{-1} such that $\text{val}_\mathbb{C}(Y^{-1}) = \text{val}_\mathbb{C}(Y)^{-1} = \text{uval}_{\mathbb{A}_j}(Y)^{-1}$ and $\text{val}_\mathbb{C}(Z^{-1}) = \text{val}_\mathbb{C}(Z)^{-1} = \text{uval}_{\mathbb{A}_j}(Z)^{-1}$. In case $Y \in \text{Lo}_j$ (respectively, $Z \in \text{Lo}_j$), the symbol Y^{-1} (respectively, Z^{-1}) is already present (since $\mathbb{A}_j$ is saturated).

Now we set $\Gamma_i = \mathsf{Lo}_j(i)$ and apply Proposition 5.35 (and the analogous statement for $\backslash_s$ and $\sqcap_s$) to Γ_i ($i \in W$) and $\mathbb{C}$ to obtain two PC-expressions α and β such that $|\alpha|, |\beta| \le \mathscr{O}(|\mathbb{A}|)$, $\mathsf{Var}(\alpha) = \{Y\}$, $\mathsf{Var}(\beta) = \{Z\}$, and

$$[\mathsf{val}_{\mathbb{C}}(\alpha)]_{I_j} = [\mathsf{val}_{\mathbb{C}}(Y)]_{I_j} \backslash_s [\mathsf{val}_{\mathbb{C}}(Z^{-1})]_{I_j},$$

$$[\mathsf{val}_{\mathbb{C}}(\beta)]_{I_j} = [\mathsf{val}_{\mathbb{C}}(Z)]_{I_j} \backslash_p [\mathsf{val}_{\mathbb{C}}(Y^{-1})]_{I_j}.$$

But due to the correspondence between $\mathbb{C}$ and $\mathsf{up}(\mathbb{A}_j)$, this means (uval denotes $\mathsf{uval}_{\mathbb{A}_j}$)

$$[\mathsf{uval}(\alpha)]_{I_j} = [\mathsf{uval}(Y)]_{I_j} \backslash_s [\mathsf{uval}(Z)]_{I_j}^{-1},$$

$$[\mathsf{uval}(\beta)]_{I_j} = [\mathsf{uval}(Z)]_{I_j} \backslash_p [\mathsf{uval}(Y)]_{I_j}^{-1}.$$

Moreover, for every subexpression of the form $\pi_\Delta(\gamma)$ in α or β, there exists $K \subseteq \{1, \ldots, n\}$ with $\Delta = \bigcup_{i \in K} \mathsf{Lo}_j(i)$. Intuitively, α and β represent the parts of $\mathsf{uval}(Y)$ and $\mathsf{uval}(Z)$ that remain after cancellation in the graph group generated by the alphabet Lo_j. Hence, $[\mathsf{uval}(\alpha)]_{I_j} [\mathsf{uval}(\beta)]_{I_j}$ does not contain a factor of the form $[XX^{-1}]_{I_j}$ for $X \in \mathsf{Lo}_j$. Using Lemma 5.26 we can compute the sets $V_{\max} = \max([\mathsf{uval}(\alpha)]_{I_j})$ and $V_{\min} = \min([\mathsf{uval}(\beta)]_{I_j})$ in polynomial time. In order to apply Lemma 5.26 we have to compute, using Lemma 3.16, temporary SLPs for the words $\mathsf{uval}(\alpha)$ and $\mathsf{uval}(\beta)$. These SLPs are temporary in the sense that they are not needed for the next iteration of the algorithm. Recall that $V_{\max}$ and $V_{\min}$ are subsets of Lo_j. Since every independence clique of (Lo_j, I_j) has size at most $|W| = n = \mathscr{O}(1)$, we have $|V_{\max}|, |V_{\min}| = \mathscr{O}(1)$.

Next, Lemma 5.27 allows us to compute in polynomial time PC-expressions α', β' such that $\mathsf{Var}(\alpha') = \mathsf{Var}(\alpha) = \{Y\}$, $\mathsf{Var}(\beta') = \mathsf{Var}(\beta) = \{Z\}$, and

$$[\mathsf{uval}(\alpha')]_{I_j} = [\mathsf{uval}(\alpha)]_{I_j} \backslash_s [V_{\max}]_{I_j},$$

$$[\mathsf{uval}(\beta')]_{I_j} = [\mathsf{uval}(\beta)]_{I_j} \backslash_p [V_{\min}]_{I_j}.$$

The length bound in Lemma 5.27 implies that $|\alpha'|, |\beta'| \le \mathscr{O}(|\mathbb{A}|)$. Moreover, for every $1 \le i \le n$ we must have $|V_{\max} \cap \mathsf{Lo}_j(i)| \le 1$ and $|V_{\min} \cap \mathsf{Lo}_j(i)| \le 1$. Let

$$V'_{\max} = \{X \in V_{\max} \mid \mathsf{type}_j(X) \notin \mathsf{type}_j(V_{\min})\},$$

$$V'_{\min} = \{X \in V_{\min} \mid \mathsf{type}_j(X) \notin \mathsf{type}_j(V_{\max})\}.$$

If $(X_1, X_2) \in V_{\max} \times V_{\min}$ is such that $\mathsf{type}_j(X_1) = \mathsf{type}_j(X_2) = i$, then by the definition of $[\mathsf{uval}(\alpha)]_{I_j}$ and $[\mathsf{uval}(\beta)]_{I_j}$, we must have $\mathsf{val}(X_1)\mathsf{val}(X_2) \ne 1$ in G_i (no further cancellation is possible in the product $[\mathsf{uval}(\alpha)]_{I_j} [\mathsf{uval}(\beta)]_{I_j}$). For each such pair we add a new lower level variable X_{X_1, X_2} to Lo_j with right-hand side $X_1 X_2$; let V' be the set of these new variables. Clearly, $|V'| \le n = \mathscr{O}(1)$. Finally, the right-hand side for X is changed to the PC-expression

$$\gamma = \alpha' v'_{\max} v' v'_{\min} \beta', \tag{5.3}$$

where $v'_{\max}$ (respectively, v', $v'_{\min}$) is an arbitrary word that enumerates all variables from $V'_{\max}$ (respectively, V', $V'_{\min}$). We have $|\gamma| = |\alpha'| + |\beta'| + \mathcal{O}(1) \leq \mathcal{O}(|\mathbb{A}|)$. Clearly, γ evaluates in G to the same group element as $\mathsf{val}_{\mathbb{A}_j}(X)$. By adding at most $|V'| = \mathcal{O}(1)$ many further lower level variables, we obtain a saturated system. The resulting 2-level PCSLP is not necessarily irredundant, but this can be ensured, as explained in the paragraph after Definition 5.30, using oracle calls to the compressed word problems for the vertex groups G_i (this does not increase the size of the 2-level PCSLP). The resulting 2-level PCSLP is pure, irredundant, and saturated, but not necessarily nicely projecting, because of the new lower level variables from V'. But note that these variables do not occur in the scope of a projection operator π_Δ; they only occur in the words v' in (5.3). Hence, we may add the new lower level variables to the appropriate sets appearing in projection operators, so that the 2-level PCSLP becomes nicely projecting as well. The resulting 2-level PCSLP is $\mathbb{A}_{j+1}$; it is well formed. Its size can be bounded by $|\mathbb{A}_j| + \mathcal{O}(|\mathbb{A}|) \leq 2 \cdot |\mathbb{A}| + \mathcal{O}(j \cdot |\mathbb{A}|) + \mathcal{O}(|\mathbb{A}|) \leq 2 \cdot |\mathbb{A}| + \mathcal{O}((j + 1) \cdot |\mathbb{A}|)$; hence, (a) and (b) above hold for $\mathbb{A}_{j+1}$. Moreover, in the group G we have $\mathsf{val}(\mathbb{A}_{j+1}) = \mathsf{val}(\mathbb{A}_j) = \mathsf{val}(\mathbb{A})$; hence, (c) holds. Lemma 5.23 implies that X is reduced in $\mathbb{A}_{j+1}$ which implies property (d) for $\mathbb{A}_{j+1}$. Finally, for (e) note that $|\mathsf{uval}(\mathbb{A}_{j+1})| \leq |\mathsf{uval}(\mathbb{A}_j)| \leq |\mathsf{val}(\mathbb{A})|$.

After $r \leq |V_\mathbb{A}|$ steps, our construction yields the well-formed and reduced 2-level PCSLP $\mathbb{A}_r$ with $\mathsf{val}(\mathbb{A}_r) = \mathsf{val}(\mathbb{A})$ in G. This proves Proposition 5.34. $\qquad\square$

We can now easily finish the proof of Theorem 5.4.

Proof of Theorem 5.4. Let $\mathbb{A}$ be an SLP over the monoid generating set $\Sigma \cup \Sigma^{-1}$ of the graph product G. By Proposition 5.34 we can translate $\mathbb{A}$ into a reduced and well-formed 2-level PCSLP $\mathbb{B}$ with $\mathsf{val}(\mathbb{A}) = \mathsf{val}(\mathbb{B})$ in G. This translation can be done in polynomial time using oracle access to the problems $\mathsf{CWP}(G_i)$ for $i \in W$. By Lemma 5.33, we have $\mathsf{val}(\mathbb{A}) = 1$ in G if and only if $\mathsf{uval}(\mathbb{B}) = \varepsilon$. By Lemma 3.16, we can translate $\mathsf{up}(\mathbb{B})$ in polynomial time into an equivalent SLP, for which it is trivial to check whether it produces the empty word. This proves Theorem 5.4. $\qquad\square$

5.3.1 Proof of Proposition 5.35

We will prove Proposition 5.35 in this section. Recall that we fixed the finite undirected graph (W, E) with $W = \{1, \dots, n\}$.

Proposition 5.35 (restated). *Let (W, E) be a fixed independence alphabet with $W = \{1, \dots, n\}$. The following problem can be solved in polynomial time:*

input: Pairwise disjoint finite alphabets $\Gamma_1, \dots, \Gamma_n$, an SLP $\mathbb{B}$ over the terminal alphabet $\Gamma = \bigcup_{i=1}^n \Gamma_i$, and two variables Y and Z from $\mathbb{B}$.
output: PC-expressions α, β with $\mathsf{Var}(\alpha) = \mathsf{Var}(\beta) = \{Y\}$ such that the following holds, where $I = E[\Gamma_1, \dots, \Gamma_n]$:

(1) *For every subexpression of the form* $\pi_\Delta(\gamma)$ *in* α *or* β*, there exists* $K \subseteq W$ *with*
   $\Delta = \bigcup_{i \in K} \Gamma_i$.

(2) $[\mathsf{val}_\mathbb{B}(\alpha)]_I = [\mathsf{val}_\mathbb{B}(Y)]_I \setminus_p [\mathsf{val}_\mathbb{B}(Z)]_I$

(3) $[\mathsf{val}_\mathbb{B}(\beta)]_I = [\mathsf{val}_\mathbb{B}(Y)]_I \sqcap_p [\mathsf{val}_\mathbb{B}(Z)]_I$

(4) $|\alpha|, |\beta| \leq \mathscr{O}(\log |\mathsf{val}(\mathbb{B})|)$

In the following, we will write $\setminus$ and $\sqcap$ for $\setminus_p$ and $\sqcap_p$, respectively. Moreover, if $\Delta = \bigcup_{i \in K} \Gamma_i$, we will write π_K for the projection morphism $\pi_\Delta : \Gamma^* \to \Gamma^*$. Let us fix $\Gamma = \bigcup_{i=1}^{n} \Gamma_i$ and let $I = E[\Gamma_1, \dots, \Gamma_n]$.

Let $s \in \Gamma^*$ be a word and $J \subseteq \{1, \dots, |s|\}$ a set of positions in s. Below, we identify the dependence graph D_s with the edge relation of D_s. We are looking for a succinct representation for the set of all positions p such that $\exists j \in J : (j, p) \in D_s^*$. This is the set of positions p for which there exists a path in the dependence graph D_s from some position $j \in J$ to position p. For $i \in W$ define

$$\mathsf{pos}(s, J, i) = \min(\{|s| + 1\} \cup \{p \mid 1 \leq p \leq |s|, s[p] \in \Gamma_i, \exists j \in J : (j, p) \in D_s^*\}).$$

Example 5.36. To ease the reading we will consider the set $W = \{a, b, c, d, e\}$ instead of $W = \{1, \dots, 5\}$. The dependence relation D is

$$
\begin{array}{c}
a \qquad e \\
\diagdown \quad \diagup \diagdown \\
b \rule{1cm}{0.4pt} c \rule{1cm}{0.4pt} d
\end{array}
$$

Let $\Gamma_x = \{x\}$ for $x \in W$. We consider the following word s, where we write the position number on top of each symbol:

$$
\begin{array}{cccccccccccccccc}
1 & 2 & 3 & 4 & 5 & 6 & 7 & 8 & 9 & 10 & 11 & 12 & 13 & 14 & 15 \\
s = d & b & c & d & b & a & c & d & b & d & e & a & b & d & c
\end{array}
$$

The dependence graph of s looks as follows:

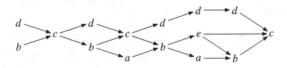

Let $J = \{5, 6, 9\}$. We want to determine $\mathsf{pos}(s, J, 4)$. In the following picture of the dependence graph of s we mark positions from J with boxes and all positions $p \notin J$ with $(j, p) \in D_s^*$ for some $j \in J$ with circles.

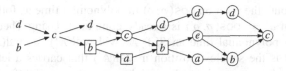

The positions with letters from $\Gamma_4 = \{d\}$ which depend from positions from J are $\{8, 10, 14\}$ with the minimum 8; hence, $\mathsf{pos}(s, J, 4) = 8$.

For the set $J = \{6, 9\}$ we get the following picture for $\mathsf{pos}(s, J, 4)$:

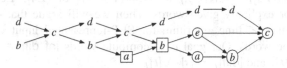

Since there are no positions with letters from $\Gamma_4 = \{d\}$ which depend from positions from J, it follows that $\mathsf{pos}(s, J, 4) = |s| + 1 = 16$.

Instead of $\mathsf{pos}(s, \{p\}, i)$, we simply write $\mathsf{pos}(s, p, i)$. Note that $\mathsf{pos}(s, \emptyset, i) = |s| + 1$. The definition of $\mathsf{pos}(s, J, i)$ and the fact that symbols from a set Γ_i are pairwise dependent implies

Lemma 5.37. *Let $s \in \Gamma^*$ and $J \subseteq \{1, \ldots, |s|\}$. Then for every position $1 \le p \le |s|$ the following two properties are equivalent:*

- $\exists j \in J : (j, p) \in D_s^*$.
- *If $s[p] \in \Gamma_i$, then $p \ge \mathsf{pos}(s, J, i)$.*

We will also need the following lemma.

Lemma 5.38. *For a given SLP $\mathbb{A}$, a position $1 \le p \le |\mathsf{val}(\mathbb{A})|$, and $i \in W$, we can compute the position $\mathsf{pos}(\mathsf{val}(\mathbb{A}), p, i)$ in polynomial time.*

Proof. We first need a few definitions: Let $D = (W \times W) \setminus E$ be the dependence relation for our fixed independence alphabet (W, E). A path in (W, D) (viewed as an undirected graph) is called *simple*, if it does not visit a node twice. For $j \in W$ let $\mathscr{P}_j$ be the set of all simple paths in the dependence alphabet (W, D) that start in the node j. The path, which only consists of the node j, belongs to $\mathscr{P}_j$. The size of each set $\mathscr{P}_j$ is bounded by $2^{|W|} |W|!$, which is a fixed constant in our consideration.

Let us now fix $\mathbb{A}$, p, and i as in the lemma. By Proposition 3.9(3), we can compute in polynomial time the unique node $j \in W$ such that $\mathsf{val}(\mathbb{A})[p] \in \Gamma_j$. For a simple path $\rho \in \mathscr{P}_j$ let us define $\mathsf{pos}(p, \rho) \in \{1, \ldots, |\mathsf{val}(\mathbb{A})| + 1\}$ inductively. Let $\rho = (j_1, j_2, \ldots, j_\ell)$ with $j = j_1$. If $\ell = 1$, then $\mathsf{pos}(p, \rho) = p$. Otherwise, let $\rho' = (j_1, j_2, \ldots, j_{\ell-1})$ and let $\mathsf{pos}(p, \rho)$ be the smallest position $q > \mathsf{pos}(p, \rho')$ such that $\mathsf{val}(\mathbb{A})[q] \in \Gamma_{j_\ell}$ if such a position exists; otherwise, $\mathsf{pos}(p, \rho) = |\mathsf{val}(\mathbb{A})| + 1$.

We can compute the position $\mathsf{pos}(p, \rho)$ in polynomial time as follows: Assume that the position $q' = \mathsf{pos}(p, \rho')$ is already computed. Using Theorem 3.14 we can compute in polynomial time an SLP $\mathbb{A}'$ with $\mathsf{val}(\mathbb{A}') = \mathsf{val}(\mathbb{A})[q' + 1 :]$. Then $\mathsf{pos}(p, \rho)$ is the smallest position in $\mathsf{val}(\mathbb{A}')$ that carries a letter from Γ_{j_ℓ}. This position can be computed in polynomial time using a straightforward variation of Proposition 3.9(5). This shows that the number $\mathsf{pos}(p, \rho)$ can be computed in polynomial time for every simple path $\rho \in \mathscr{P}_j$. Finally, $\mathsf{pos}(\mathsf{val}(\mathbb{A}), p, i)$ is the minimum over all these positions for all simple paths from j to i.   □

Let us now come back to the problem of constructing PC-expressions, which evaluates to $[\mathsf{val}_\mathbb{B}(Y)]_I \setminus [\mathsf{val}_\mathbb{B}(Z)]_I$ and $[\mathsf{val}_\mathbb{B}(Y)]_I \sqcap [\mathsf{val}_\mathbb{B}(Z)]_I$. Let us first solve this problem for explicitly given words. Then we will argue that our algorithm leads to a polynomial time algorithm for SLP-represented input words. Hence, let $s, t \in \Gamma^*$ be words. Our goal is to compute words $\mathsf{inf}, \mathsf{diff} \in \Gamma^*$ such that $[\mathsf{inf}]_I = [s]_I \sqcap [t]_I$ and $[\mathsf{diff}]_I = [s]_I \setminus [t]_I$.

In Algorithm 6 (Compute-Infimum-Difference) we accumulate the words inf and diff by determining for every position from $\{1, \dots, |s|\}$ (viewed as a node of the dependence graph D_s) whether it belongs to $[\mathsf{inf}]_I$ or $[\mathsf{diff}]_I$. For this, we will store a current position ℓ in the word s, which will increase during the computation. Initially, we set $\ell := 1$ and $\mathsf{inf} := \varepsilon$, $\mathsf{diff} := \varepsilon$. At the end, we have $[\mathsf{inf}]_I = [s]_I \sqcap [t]_I$ and $[\mathsf{diff}]_I = [s]_I \setminus [t]_I$.

For a set of positions $K = \{\ell_1, \dots, \ell_k\} \subseteq \{1, \dots, |s|\}$ with $\ell_1 < \ell_2 < \cdots < \ell_k$, we define the word

$$s \restriction K = s[\ell_1] \cdots s[\ell_k].$$

Consider a specific iteration of the while-loop body in algorithm Compute-Infimum-Difference and let ℓ denote the value of the corresponding program variable at the beginning of the iteration. Assume in the following that $\mathsf{Diff}_\ell \subseteq \{1, \dots, \ell - 1\}$ is the set of all positions from $\{1, \dots, \ell - 1\}$, which belong to the difference $[s]_I \setminus [t]_I$, i.e., they do not belong to the common prefix $[s]_I \sqcap [t]_I$. Moreover, let

$$\mathsf{Inf}_\ell = \{1, \dots, \ell - 1\} \setminus \mathsf{Diff}_\ell$$

be the set of all positions from $\{1, \dots, \ell - 1\}$, which belong to the trace prefix $[s]_I \sqcap [t]_I$. Thus, Inf_ℓ is downward-closed in D_s, which means that $(i, j) \in D_s^*$ and $j \in \mathsf{Inf}_\ell$ imply $i \in \mathsf{Inf}_\ell$. Moreover, we have $[s \restriction \mathsf{Inf}_\ell]_I \preceq [s]_I \sqcap [t]_I$. Note that the algorithm stores neither the set Diff_ℓ nor the set Inf_ℓ. This will be important later, when the input words s and t are represented by SLPs, because the sets Diff_ℓ and Inf_ℓ may be of exponential size in that case. If $\ell, \mathsf{inf}, \mathsf{diff}$, and $\mathsf{pos}(i)$ ($i \in W$) denote the values of the corresponding program variables at the beginning of the current iteration of the while-loop, then the algorithm will maintain the following two invariants:

(I1) $\mathsf{inf} = s \restriction \mathsf{Inf}_\ell$, $\mathsf{diff} = s \restriction \mathsf{Diff}_\ell$,
(I2) $\mathsf{pos}(i) = \mathsf{pos}(s, \mathsf{Diff}_\ell, i)$ for all $i \in W$.

Algorithm 6: Compute-Infimum-Difference

---

input : words $s, t \in \Gamma^*$
output: words inf, diff $\in \Gamma^*$ such that $[\text{inf}]_I = [s]_I \sqcap [t]_I$ and $[\text{diff}]_I = [s]_I \setminus [t]_I$
$\ell := 1$ (stores a position in s)
$\text{inf} := \varepsilon$ (stores a word)
$\text{diff} := \varepsilon$ (stores a word)
for $i \in W$ **do**
  | $\text{pos}(i) := |s| + 1$ (stores positions in s)
end
while $\ell \leq |s|$ **do**
  | $U := \{i \in W \mid \text{pos}(i) < \ell\}$
  | $\text{next} := \min(\{\text{pos}(i) \mid i \in W \setminus U\} \cup \{|s| + 1\});$
  | $j := \max\{i \mid \ell - 1 \leq i \leq \text{next} - 1, [\inf \pi_{W \setminus U}(s[\ell : i])]_I \preceq [t]_I\}$ (⋆)
  | $\text{inf} := \inf \pi_{W \setminus U}(s[\ell : j])$
  | $\text{diff} := \text{diff}\, \pi_U(s[\ell : j])\, s[j + 1]$ (let $s[|s| + 1] = \varepsilon$)
  | **for** $i \in W$ **do**
  | | $\text{pos}(i) := \min\{\text{pos}(i), \text{pos}(s, j + 1, i)\}$ (let $\text{pos}(s, |s| + 1, i) = |s| + 1$)
  | **end**
  | $\ell := j + 2$
end

In each iteration of the while-loop, we investigate the subword of s from position ℓ to the next position of the form $\text{pos}(i)$, and we determine for each position from some initial segment of this interval whether it belongs to $[s]_I \sqcap [t]_I$ or $[s]_I \setminus [t]_I$. More precisely, we search for the largest position $j \in \{\ell - 1, \ldots, \text{next} - 1\}$ such that $[\inf \pi_{W \setminus U}(s[\ell : j])]_I$ is a trace prefix of $[t]_I$. Recall that $\text{inf} = s \upharpoonright \text{Inf}_\ell$ is the already collected part of the common trace prefix. We update inf and diff by $\text{inf} := \inf \pi_{W \setminus U}(s[\ell : j])$ and $\text{diff} := \text{diff}\, \pi_U(s[\ell : j])s[j + 1]$.

Before we prove that the algorithm indeed preserves the invariants (I1) and (I2), let us first consider a detailed example.

Example 5.39. To ease the reading we will consider the set $W = \{a, b, c, d, e, f, g\}$ instead of $W = \{1, \ldots, 7\}$ together with the following dependence relation D:

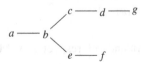

Let $\Gamma_x = \{x\}$ for all $x \in W$. We consider the following words:

$$
\begin{array}{ccccccccccccccccc}
1 & 2 & 3 & 4 & 5 & 6 & 7 & 8 & 9 & 10 & 11 & 12 & 13 & 14 & 15 & 16 & 17 \\
s = f & b & g & c & c & g & b & c & c & e & a & g & f & e & f & d & g \\
t = b & c & g & c & f & g & b & e & a & g & g & f & e & d & f & b & g \\
\end{array}
$$

The dependence graphs of $[s]_I$ and $[t]_I$ look as follows:

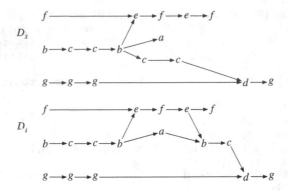

We want to determine $s \sqcap_p t$ and $s \setminus_p t$ using algorithm Compute-Infimum-Difference. Initially, we set $\ell = 1$, inf $=$ diff $= \varepsilon$, and $\mathsf{pos}(x) = |s| + 1 = 18$ for all $x \in W$. Since $\ell \le |s|$ the while-loop is executed.
First iteration: The algorithm first sets

$$U = \emptyset \quad \text{and} \quad \mathsf{next} = 18.$$

Hence, we have

$$\inf \pi_{W \setminus U}(s[\ell : \mathsf{next}-1]) = \quad |f \ \ b \ \ g \ \ c \ \ c \ \ g \ \ b \ \ c \ \ c \ \ e \ \ a \ \ g \ \ f \ \ e \ \ f \ \ d \ \ g$$
$$1 \ \ 2 \ \ 3 \ \ 4 \ \ 5 \ \ 6 \ \ 7 \ \ 8 \ \ 9 \ \ 10 \ 11 \ 12 \ 13 \ 14 \ 15 \ 16 \ 17.$$

Here, we denote with "|" the position between inf and $\pi_{W \setminus U}(s[\ell : \mathsf{next} - 1])$. The algorithm computes the largest number $0 \le j \le 17$ such that $[\inf \pi_{W \setminus U}(s[\ell : j])]_I$ is a trace prefix of $[t]_I$. From the dependence graphs above it can be easily seen that $j = 7$. We have

$$\inf \pi_{W \setminus U}(s[\ell : j]) \ = \ |f \ \ b \ \ g \ \ c \ \ c \ \ g \ \ b$$
$$\mathsf{diff} \ \pi_U(s[\ell : j])s[j+1] \ = \ \qquad\qquad\qquad\qquad c,$$

which are the new values for inf and diff, respectively. Moreover, the pos-values are reset as follows:

$$\mathsf{pos}(a) = \mathsf{pos}(b) = 18, \ \mathsf{pos}(c) = 8, \ \mathsf{pos}(d) = 16,$$
$$\mathsf{pos}(e) = \mathsf{pos}(f) = 18, \ \mathsf{pos}(g) = 17.$$

Finally, ℓ is set to 9. Since $\ell = 9 \le |s|$ the while-loop is repeated.
Second iteration: The algorithm first sets

$$U = \{c\} \quad \text{and} \quad \text{next} = 16.$$

We have

$$\inf \pi_{W \setminus \{c\}}(s[9:15]) = \begin{array}{cccccccccccccccc} f & b & g & c & c & g & b & | & e & a & g & f & e & f \\ 1 & 2 & 3 & 4 & 5 & 6 & 7 & 8 & 9 & 10 & 11 & 12 & 13 & 14 & 15. \end{array}$$

Searching for the largest position $\ell - 1 = 8 \le j \le 15 = \text{next} - 1$ such that the trace $[\inf \pi_{W \setminus U}(s[\ell : j])]_I$ is a prefix of $[t]_I$ gives $j = 15$. We have

$$\begin{array}{l} \inf \pi_{W \setminus \{c\}}(s[9:15]) = f \ b \ g \ c \ c \ g \ b \quad | \quad e \ a \ g \ f \ e \ f \\ \text{diff } \pi_{\{c\}}(s[9:15])s[16] = \qquad\qquad\quad c \ c \qquad\qquad\qquad\qquad d, \end{array}$$

which are the new values for inf and diff. The pos-values do not change in the second iteration, i.e., we still have

$$\text{pos}(a) = \text{pos}(b) = 18, \ \text{pos}(c) = 8, \ \text{pos}(d) = 16,$$

$$\text{pos}(e) = \text{pos}(f) = 18, \ \text{pos}(g) = 17.$$

Finally, ℓ is set to 17. Since $\ell \le |s|$ the while-loop is repeated.
Third iteration: The algorithm first sets

$$U = \{c, d\} \quad \text{and} \quad \text{next} = 17.$$

We have

$$\inf \pi_{W \setminus U}(s[17:16]) = \begin{array}{cccccccccccccccc} f & b & g & c & c & g & b & & e & a & g & f & e & f| \\ 1 & 2 & 3 & 4 & 5 & 6 & 7 & 8 & 9 & 10 & 11 & 12 & 13 & 14 & 15 & 16. \end{array}$$

We find $j = 16$. We have

$$\begin{array}{l} \inf \pi_{W \setminus U}(s[17:16]) = f \ b \ g \ c \ c \ g \ b \qquad e \ a \ g \ f \ e \ f| \\ \text{diff } \pi_U(s[17:16])s[17] = \qquad\qquad\quad c \ c \qquad\qquad\qquad\quad d \ g. \end{array}$$

Also in the third iteration, the pos-values do not change. Finally, ℓ is set to 18. Since $\ell > |s|$ the algorithm stops and produces $[\inf]_I = [fbgccgbeagfef]_I$ and

$[\mathrm{diff}]_I = [ccdg]_I$. These traces are indeed $[s]_I \sqcap [t]_I$ and $[s]_I \setminus [t]_I$. This is visualized in the next picture with $[s]_I \sqcap [t]_I$ on the left side of the dotted line and $[s]_I \setminus [t]_I$ on the right side.

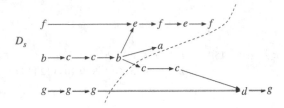

Let us now prove the correctness of the algorithm. We start with invariant (I1).

Lemma 5.40. *Algorithm Compute-Infimum-Difference preserves invariant (I1).*

Proof. Let us take $\ell \in \{1, \dots, |s|\}$ and assume that the invariants (I1) and (I2) hold at the beginning of an execution of the body of the while-loop. Hence,

- $\mathsf{inf} = s \restriction \mathsf{Inf}_\ell$, $\mathsf{diff} = s \restriction \mathsf{Diff}_\ell$, and
- $\mathsf{pos}(i) = \mathsf{pos}(s, \mathsf{Diff}_\ell, i)$ for all $i \in W$.

We have to show that invariant (I1) holds after the execution of the loop body as well. As in the algorithm, let:

$$U = \{i \in W \mid \mathsf{pos}(s, \mathsf{Diff}_\ell, i) < \ell\}, \tag{5.4}$$

$$\mathsf{next} = \min(\{\mathsf{pos}(s, \mathsf{Diff}_\ell, i) \mid i \in W \setminus U\} \cup \{|s| + 1\}), \tag{5.5}$$

$$j = \max\{i \mid \ell - 1 \le i \le \mathsf{next} - 1, [\mathsf{inf}\, \pi_{W \setminus U}(s[\ell : i])]_I \preceq [t]_I\}. \tag{5.6}$$

We have to prove the following statements:

- A position $p \in \{\ell, \dots, j\}$ belongs to the common trace prefix $[s]_I \sqcap [t]_I$ if and only if $s[p] \in \Gamma_i$ for some $i \in W \setminus U$.
- If $j + 1 \le |s|$, then $j + 1$ does not belong to the common trace prefix $[s]_I \sqcap [t]_I$.

For the first point, assume that $s[p] \in \Gamma_i$, where $\ell \le p \le j$ and $i \in U$. By definition of U in (5.4), we have $\mathsf{pos}(s, \mathsf{Diff}_\ell, i) < \ell \le p$. Lemma 5.37 implies that there exists a path in D_s from some position in Diff_ℓ to position p. Since positions in Diff_ℓ do not belong to $[s]_I \sqcap [t]_I$, position p does not belong to $[s]_I \sqcap [t]_I$ as well.

For the other direction, consider the set of positions

$$P = \{p \mid \ell \le p \le j, s[p] \in \Gamma_i \text{ for some } i \in W \setminus U\}.$$

We claim that $\mathsf{Inf}_\ell \cup P$ is a downward-closed subset in D_s. Since $[s \upharpoonright (\mathsf{Inf}_\ell \cup P)]_I = [\inf \pi_{W \setminus U}(s[\ell : j])]_I \preceq [t]_I$ by (5.6), this implies that all positions from P indeed belong to $[s]_I \sqcap [t]_I$. That $\mathsf{Inf}_\ell \cup P$ is downward-closed in D_s follows from the following three facts:

- Inf_ℓ is downward-closed in D_s.
- There does not exist a path from a node in Diff_ℓ to a node from P: Assume that such a path, ending in $p \in P$, would exist. Let $s[p] \in \Gamma_i$ with $i \in W \setminus U$. Lemma 5.37 implies $\mathsf{pos}(s, \mathsf{Diff}_\ell, i) \leq p$. Moreover, $i \in W \setminus U$ implies $\mathsf{pos}(s, \mathsf{Diff}_\ell, i) \geq \ell$ by (5.4). Hence, $\ell \leq \mathsf{pos}(s, \mathsf{Diff}_\ell, i) \leq p \leq j < \mathsf{next}$, where the last inequality follows from (5.6). But this contradicts the definition of next in (5.5).
- There does not exist a path from a node in $\{\ell, \ldots, j\} \setminus P$ to a node of P: We have $\{\ell, \ldots, j\} \setminus P = \{p \mid \ell \leq p \leq j, s[p] \in \Gamma_i \text{ for some } i \in U\}$. Lemma 5.37 and the definition of U from (5.4) imply that every node from $\{\ell, \ldots, j\} \setminus P$ can be reached via a path starting in Diff_ℓ. Hence, the existence of a path from $\{\ell, \ldots, j\} \setminus P$ to P contradicts the previous point.

It remains to be shown that position $j + 1$ does not belong to the common trace prefix $[s]_I \sqcap [t]_I$ in case $j + 1 \leq |s|$. We distinguish several cases: If $j = \mathsf{next} - 1$, then $j + 1 = \mathsf{next} = \mathsf{pos}(s, \mathsf{Diff}_\ell, i)$ for some $i \in W \setminus U$ by (5.5). Hence, there exists a path from Diff_ℓ to $j + 1$ in D_s. Therefore $j + 1$ cannot belong to $[s]_I \sqcap [t]_I$. Now, assume that $j < \mathsf{next} - 1$. If $s[j + 1] \in \Gamma_i$ for some $i \in U$, then (5.4) implies $\mathsf{pos}(s, \mathsf{Diff}_\ell, i) < \ell \leq j + 1$. Lemma 5.37 again yields the existence of a path from Diff_ℓ to $j + 1$. Finally, let $s[j + 1] \in \Gamma_i$ for some $i \in W \setminus U$. Maximality of j in (5.6) implies that the trace

$$[\inf \pi_{W \setminus U}(s[\ell, j + 1])]_I = [\inf \pi_{W \setminus U}(s[\ell : j])]_I s[j + 1]$$

is not a trace prefix of $[t]_I$. Since we already know that the trace $[\inf \pi_{W \setminus U}(s[\ell : j])]_I$ consists exactly of those positions from $\{1, \ldots, j\}$ that belong to the common trace prefix $[s]_I \sqcap [t]_I$, this implies that $j + 1$ does not belong to $[s]_I \sqcap [t]_I$. $\square$

Lemma 5.41. *Algorithm Compute-Infimum-Difference preserves invariant (I2).*

Proof. We consider a specific iteration of the while-loop and assume that (I1) and (I2) hold at the beginning of the loop, i.e.,

- $\mathsf{inf} = s \upharpoonright \mathsf{Inf}_\ell$, $\mathsf{diff} = s \upharpoonright \mathsf{Diff}_\ell$ and
- $\mathsf{pos}(i) = \mathsf{pos}(s, \mathsf{Diff}_\ell, i)$ for all $i \in W$.

We infer that (I2) holds after the execution of the loop. Let U, next, and j be defined by (5.4)–(5.6). Let $\ell' = j + 2 > \ell$ be the new value of ℓ after the execution of the loop body and let $i \in W$. We have to show that $\mathsf{pos}(s, \mathsf{Diff}_{\ell'}, i)$ is the new value of $\mathsf{pos}(i)$ after the execution of the loop body. Since we already know by Lemma 5.40 that (I1) holds after the execution of the loop, we have

$$s \upharpoonright \mathsf{Diff}_{\ell'} = \mathsf{diff}\, \pi_U(s[\ell : j])\, s[j + 1] = (s \upharpoonright \mathsf{Diff}_\ell)\, \pi_U(s[\ell : j])\, s[j + 1],$$

Algorithm 7: Compute-Compressed-Infimum-Difference

---
input : SLP $\mathbb{B}$ and two variables Y and Z
output: PC-expressions α, β with $\mathsf{val}(\alpha) = [\mathsf{val}(Y)]_I \sqcap [\mathsf{val}(Z)]_I$ and
$\qquad \mathsf{val}(\beta) = [\mathsf{val}(Y)]_I \setminus [\mathsf{val}(Z)]_I$
$\ell := 1$
$\alpha := \varepsilon$
$\beta := \varepsilon$
for $i \in W$ **do**
$\quad | \quad \mathsf{pos}(i) := |\mathsf{val}(Y)| + 1$
end
while $\ell \leq |\mathsf{val}(Y)|$ **do**
$\quad | \quad U := \{i \in W \mid \mathsf{pos}(i) < \ell\}$
$\quad | \quad \mathsf{next} := \min(\{\mathsf{pos}(i) \mid i \in W \setminus U\} \cup \{|\mathsf{val}(Y)| + 1\})$
$\quad | \quad j := \max\{i \mid \ell - 1 \leq i \leq \mathsf{next} - 1, [\mathsf{val}(\alpha \circ \pi_{W \setminus U}(Y[\ell : i]))]_I \preceq [\mathsf{val}(Z)]_I\}$ $\qquad (\star)$
$\quad | \quad \alpha := \alpha \circ \pi_{W \setminus U}(Y[\ell : j])$
$\quad | \quad \beta := \beta \circ \pi_U(Y[\ell : j]) \circ Y[j + 1]$ $\qquad\qquad$ (let $\mathsf{val}(Y)[|\mathsf{val}(Y)| + 1] = \varepsilon$)
$\quad | \quad$ **for** $i \in W$ **do**
$\quad | \quad\quad | \quad \mathsf{pos}(i) := \min\{\mathsf{pos}(i), \mathsf{pos}(\mathsf{val}(Y), j + 1, i)\}$ $\qquad\qquad (\star\star)$
$\quad | \quad$ **end**
$\quad | \quad \ell := j + 2$
end

---

which means that

$$\mathsf{Diff}_{\ell'} = \mathsf{Diff}_\ell \cup \{p \mid \ell \leq p \leq j, \exists k \in U : s[p] \in \Gamma_k\} \cup \{j + 1\} \qquad (5.7)$$

(in case $j = |s|$, we omit $\{j + 1\}$ from the right-hand side). Hence, we have to show that

$$\mathsf{pos}(s, \mathsf{Diff}_{\ell'}, i) = \min\{\mathsf{pos}(s, \mathsf{Diff}_\ell, i), \mathsf{pos}(s, \{j + 1\}, i)\}. \qquad (5.8)$$

for all $i \in W$. By (5.7), $\mathsf{Diff}_\ell \cup \{j + 1\} \subseteq \mathsf{Diff}_{\ell'}$, which implies

$$\mathsf{pos}(s, \mathsf{Diff}_{\ell'}, i) \leq \min\{\mathsf{pos}(s, \mathsf{Diff}_\ell, i), \mathsf{pos}(s, j + 1, i)\}.$$

It remains to show that

$$\mathsf{pos}(s, \mathsf{Diff}_{\ell'}, i) \geq \min\{\mathsf{pos}(s, \mathsf{Diff}_\ell, i), \mathsf{pos}(s, j + 1, i)\}.$$

The case that $\mathsf{pos}(s, \mathsf{Diff}_{\ell'}, i) = |s| + 1$ is trivial. Hence, assume that $\mathsf{pos}(s, \mathsf{Diff}_{\ell'}, i) \leq |s|$ and consider a path in D_s from a position $p \in \mathsf{Diff}_{\ell'}$ to a position $q \leq |s|$ such that $s[q] \in \Gamma_i$. It suffices to show that there is a path from a position in $\mathsf{Diff}_\ell \cup \{j + 1\}$ to p (then, there exists a path from $\mathsf{Diff}_\ell \cup \{j + 1\}$ to q as well). By (5.7), we have

$$p \in \mathsf{Diff}_\ell \cup \{p \mid \ell \leq p \leq j, \exists k \in U : s[p] \in \Gamma_k\} \cup \{j + 1\}.$$

The case $p \in \mathsf{Diff}_\ell \cup \{j+1\}$ is trivial. Hence, assume that $\ell \le p \le j$ and $s[p] \in \Gamma_k$ for some $k \in U$. From (5.4), we get $\mathsf{pos}(s, \mathsf{Diff}_\ell, k) < \ell \le p$. Lemma 5.37 implies that there exists a path from Diff_ℓ to p. $\qquad\square$

Lemma 5.42. *The number of iterations of the while-loop in algorithm Compute-Infimum-Difference is bounded by $|W| + 1 = n + 1 = \mathcal{O}(1)$.*

Proof. We claim that in each execution of the loop body except for the last one, the set $U = \{i \in W \mid \mathsf{pos}(i) < \ell\}$ strictly grows, which proves the lemma. Let us consider an execution of the loop body. Note that for every $i \in W$, position $\mathsf{pos}(i)$ cannot increase. There are two cases to distinguish. If $j < \mathsf{next} - 1$, then the symbol $s[j+1]$ must belong to some alphabet Γ_i with $i \in W \setminus U$ due to the maximality of j in line (*) of the algorithm. Clearly, $\mathsf{pos}(s, \{j+1\}, i) = j+1$; hence, $\mathsf{pos}(i)$ will be set to a value $\le j+1$ in the loop body. Since the new value ℓ will be $j+2$, the new set U will also contain i, i.e., U strictly grows. If $j = \mathsf{next} - 1 < |s|$, then again, since $j+1 = \mathsf{next} = \mathsf{pos}(i)$ for some $i \in W \setminus U$ and the new value ℓ will be $j+2$, the set U strictly grows. Finally, if $j = \mathsf{next} - 1 = |s|$, then ℓ will be set to $|s| + 2$ and the algorithm terminates. $\qquad\square$

Algorithm Compute-Infimum-Difference for computing $[s]_I \setminus [t]_I$ and $[s]_I \sqcap [t]_I$ leads to Algorithm 7 (Compute-Compressed-Infimum-Difference), which computes PC-expressions for (we write val for $\mathsf{val}_\mathbb{B}$ in the following) $[\mathsf{val}(Y)]_I \sqcap [\mathsf{val}(Z)]_I$ and $[\mathsf{val}(Y)]_I \setminus [\mathsf{val}(Z)]_I$. For better readability we denote the concatenation operation in PC-expressions by $\circ$ in Algorithm 7. The idea is to consider the statements for updating inf and diff in algorithm Compute-Infimum-Difference as statements for computing PC-expressions α and β with $[\mathsf{val}(\alpha)]_I = [\mathsf{val}(Y)]_I \sqcap [\mathsf{val}(Z)]_I$ and $[\mathsf{val}(\beta)]_I = [\mathsf{val}(Y)]_I \setminus [\mathsf{val}(Z)]_I$. So, (2) and (3) from Proposition 5.35 are satisfied. Moreover, property (1) follows directly from the construction of α and β. For the size estimate in (4), note that by Lemma 5.42, α and β are concatenations of $\mathcal{O}(1)$ many expressions of the form $\pi_K(Y[p_1, p_2])$. Moreover, each of the positions p_1 and p_2 is bounded by $|\mathsf{val}(Y)| \le |\mathsf{val}(\mathbb{B})|$ and hence needs only $\mathcal{O}(\log |\mathsf{val}(\mathbb{B})|)$ many bits.

It remains to argue that Algorithm 7 can be implemented such that it runs in polynomial time. By Lemma 5.42, the number of iterations of the loop body is bounded by the constant $|W| + 1$. Hence, it suffices to show that a single iteration only needs polynomial time. The condition $[\mathsf{val}(\alpha \circ \pi_{W \setminus U}(Y[\ell : j]))]_I \preceq [\mathsf{val}(Z)]_I$ in line (*) of Algorithm 7 can be checked in polynomial time by Lemma 5.25. For this, note that by Lemma 3.16 we can compute in polynomial time an SLP for the word $\mathsf{val}(\alpha \circ \pi_{W \setminus U}(Y[\ell : j]))$. Hence, the number j in line (*) can be computed in polynomial time via binary search in the same way as in the proof of Theorem 4.11 for free groups. Finally, the position $\mathsf{pos}(\mathsf{val}(Y), j+1, i)$ in line (**) of Algorithm 7 can be computed in polynomial time by Lemma 5.38. This concludes the proof of Proposition 5.35 and hence the proof of Theorem 5.4, which is the main result of this chapter.

Chapter 6
The Compressed Word Problem in HNN-Extensions

In this chapter we prove two further important transfer theorems for the compressed word problem:

- The compressed word problem for a multiple HNN-extension of a base group H over finite associated subgroups is polynomial time Turing-reducible to the compressed word problem for H.
- The compressed word problem for an amalgamated product of group H_1 and H_2 with finite amalgamated subgroups is polynomial time Turing-reducible to the compressed word problems for H_1 and H_2.

"HNN" stands for the authors Higman, Neumann, and Neumann of [78], where they introduced HNN-extension. An HNN-extension is specified by a base group H and two isomorphic subgroups A and B (the associated subgroup) together with an isomorphism $\varphi : A \to B$. One would like to extend this isomorphism to an automorphism on H. But in general this is not possible. But H embeds into a group on which φ is extended to an automorphism. This group is obtained from H by adding a new generator t (the so-called stable letter) together with the relations $t^{-1}at = \varphi(a)$ for all $a \in A$. Thus, conjugation of A by t gives the isomorphism φ. This operation can be generalized to multiple HNN-extensions, where several stable letters are added in one step.

The amalgamated free product of two groups H_1 and H_2 with isomorphic subgroups A_1 and A_2 and an isomorphism $\varphi : A_1 \to A_2$ is obtained by identifying in the free product $H_1 * H_2$ the elements a and $\varphi(a)$ for all $a \in A_1$.

HNN-extension and amalgamated free products are extremely important in combinatorial group theory. For instance, finitely presented groups with an undecidable word problem (see Theorem 2.14) can be constructed from free groups using a series of HNN-extensions. Bass-Serre theory [153] relates the action of a group on a tree with an iterated decomposition of the group via HNN-extensions and amalgamated free products. By Stallings famous end theorem [156], a finitely generated group has infinitely many ends (a certain geometric invariant of a group) if and only if the group can be written as an HNN-extension with finite associated subgroups or an amalgamated free product with finite amalgamated subgroups.

M. Lohrey, *The Compressed Word Problem for Groups*, SpringerBriefs in Mathematics, 115
DOI 10.1007/978-1-4939-0748-9_6, © Markus Lohrey 2014

In contrast to general HNN-extensions and amalgamated products, the restriction to finite associated (respectively, amalgamated) subgroups is algorithmically tame. It is not too difficult to show that the word problem for an HNN-extension G with base group H and finite associated subgroups is polynomial time Turing-reducible to the word problem for H, and a similar result holds for amalgamated free products. By the results of this chapter, this tameness also holds in the compressed setting. The material of this chapter is taken from [73], which is partly inspired by the work from [117] on solving equations in HNN-extensions.

6.1 HNN-Extensions

We introduce multiple HNN-extensions in this section. Let us fix throughout this chapter a *base group* $H = \langle \Sigma \mid R \rangle$. Let us also fix isomorphic subgroups $A_i, B_i \leq H$ $(1 \leq i \leq n)$ and isomorphisms $\varphi_i : A_i \to B_i$. Let $h : (\Sigma \cup \Sigma^{-1})^* \to H$ be the canonical morphism, which maps a word $w \in (\Sigma \cup \Sigma^{-1})^*$ to the element of H it represents. Let $t_1, \ldots, t_n$ symbols which do not belong to the group H. It is common to write a^{t_i} an abbreviation for $t_i^{-1} a t_i$.

Definition 6.1 (multiple HNN-extension). The *(multiple) HNN-extension* defined by H and $A_1, B_1, \ldots, A_n, B_n$ is the group

$$G = \langle H, t_1, \ldots, t_n \mid a^{t_i} = \varphi_i(a) \ (1 \leq i \leq n, a \in A_i) \rangle. \tag{6.1}$$

The first important result about HNN-extensions is:

Theorem 6.2 (cf. [78]). *For two words $u, v \in (\Sigma \cup \Sigma^{-1})^*$ we have $u = v$ in G if and only if $u = v$ in H. Hence, H is a subgroup of G.*

In this chapter, we will be only concerned with the case that all groups $A_1, \ldots, A_n$ are finite and that Σ is finite, i.e., H is finitely generated. In this situation, we may assume that $\bigcup_{i=1}^n (A_i \cup B_i) \subseteq \Sigma$. We say that A_i and B_i are *associated subgroups* in the HNN-extension G. For the following, the notations $A_i(+1) = A_i$ and $A_i(-1) = B_i$ are useful. Note that $\varphi_i^\alpha : A_i(\alpha) \to A_i(-\alpha)$ for $\alpha \in \{+1, -1\}$.

Definition 6.3 (set $\mathrm{Red}(H, \varphi_1, \ldots, \varphi_n)$ of Britton-reduced words). We say that a word $u \in (\Sigma \cup \Sigma^{-1} \cup \{t_1, t_1^{-1}, \ldots, t_n, t_n^{-1}\})^*$ is *Britton-reduced* (after John Britton) if u does not contain a factor of the form $t_i^{-\alpha} w t_i^\alpha$ for $\alpha \in \{1, -1\}$, $w \in (\Sigma \cup \Sigma^{-1})^*$, and $h(w) \in A_i(\alpha)$. With $\mathrm{Red}(H, \varphi_1, \ldots, \varphi_n)$ we denote the set of all Britton-reduced words.

For a word $u \in (\Sigma \cup \Sigma^{-1} \cup \{t_1, t_1^{-1}, \ldots, t_n, t_n^{-1}\})^*$ let us write $\pi_t(u)$ for the projection $\pi_{\{t_1, t_1^{-1}, \ldots, t_n, t_n^{-1}\}}(u)$ of u to the alphabet $\{t_1, t_1^{-1}, \ldots, t_n, t_n^{-1}\}$; see Definition 3.1. The following lemma provides a necessary and sufficient condition for equality of Britton-reduced words in an HNN-extension, and it follows easily from [119, Theorem 2.1]:

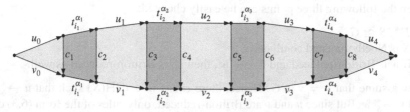

Fig. 6.1 A van Kampen diagram for the identity $u_0 t_{i_1}^{\alpha_1} u_1 t_{i_2}^{\alpha_2} u_2 t_{i_3}^{\alpha_3} u_3 t_{i_4}^{\alpha_4} u_4 = v_0 t_{j_1}^{\beta_1} v_1 t_{j_2}^{\beta_2} v_2 t_{j_3}^{\beta_3} v_3 t_{j_4}^{\beta_4} v_4$ in the HNN-extension $\langle H, t_1, \ldots, t_n \mid a^{t_i} = \varphi_i(a) \ (1 \le i \le n, a \in A_i) \rangle$. *Light-shaded* (respectively, *dark-shaded*) faces represent relations in H (respectively, relations of the form $ct_i^\alpha = t_i^\alpha \varphi_i^\alpha(c)$ with $c \in A_i(\alpha)$)

Lemma 6.4. *Let* $u = u_0 t_{i_1}^{\alpha_1} u_1 \cdots t_{i_\ell}^{\alpha_\ell} u_\ell$ *and* $v = v_0 t_{j_1}^{\beta_1} v_1 \cdots t_{j_m}^{\beta_m} v_m$ *be Britton-reduced words such that* $u_0, \ldots, u_\ell, v_0, \ldots, v_m \in (\Sigma \cup \Sigma^{-1})^*$, $\alpha_1, \ldots, \alpha_\ell, \beta_1, \ldots, \beta_m \in \{1, -1\}$, *and* $i_1, \ldots, i_\ell, j_1, \ldots, j_m \in \{1, \ldots, n\}$. *Then* $u = v$ *in the HNN-extension* G *from (6.1) if and only if the following hold:*

- $\pi_t(u) = \pi_t(v)$ *(i.e.,* $\ell = m$, $i_k = j_k$, *and* $\alpha_k = \beta_k$ *for* $1 \le k \le \ell$*)*
- *there exist* $c_1, \ldots, c_{2m} \in \bigcup_{k=1}^n (A_k \cup B_k)$ *such that:*

  - $u_k c_{2k+1} = c_{2k} v_k$ *in* H *for* $0 \le k \le \ell$ *(here we set* $c_0 = c_{2\ell+1} = 1$*)*
  - $c_{2k-1} \in A_{i_k}(\alpha_k)$ *and* $c_{2k} = \varphi_{i_k}^{\alpha_k}(c_{2k-1}) \in A_{i_k}(-\alpha_k)$ *for* $1 \le k \le \ell$.

The second condition of the lemma is visualized in Fig. 6.1 by a van Kampen diagram for uv^{-1}. The elements $c_1, \ldots, c_{2\ell}$ in such a van Kampen diagram will be also called the *connecting elements*. Note that Theorem 6.2 is an immediate consequence of Lemma 6.4 since every word $u \in (\Sigma \cup \Sigma^{-1})^*$ is Britton-reduced.

We give a proof of Lemma 6.4, following [53]:

Proof of Lemma 6.4. If a van Kampen diagram as in Fig. 6.1 exists, then clearly $u = v$ in G. For the other direction let us define the infinite alphabet $\Gamma = H \setminus \{1\} \cup \{t, t^{-1}\}$. For every $1 \le i \le n$ let X_i (resp. Y_i) be a traversal for the cosets of A_i (resp., B_i). This means that every $h \in H$ can be uniquely written as $h = ax$ (resp., $h = by$) for $a \in A_i$ and $x \in X_i$ (resp., $b \in B_i$ and $y \in Y_i$). We can assume that $1 \in X_i \cap Y_i$ for all $1 \le i \le n$. We define an infinite semi-Thue system S over Γ by the rules below, where $1 \le i \le n$, $g, h \in H \setminus \{1\}$, $a \in A_i \setminus \{1\}$, $b \in B_i \setminus \{1\}$, $x \in X_i$, and $y \in Y_i$. Moreover, for $g, h \in H \setminus \{1\}$ we denote with $g \cdot h$ either the empty word ε in case $g = h^{-1}$ or the product $gh \in H \setminus \{1\}$. Similarly, the element $1 \in X_i \cap Y_i$ is identified with the empty word.

$$t_i t_i^{-1} \to \varepsilon \qquad t_i^{-1} t_i \to \varepsilon \qquad (6.2)$$

$$gh \to g \cdot h \qquad t_i(\varphi_i(a) \cdot y) \to a t_i y \qquad t_i^{-1}(a \cdot x) \to \varphi_i(a) t_i^{-1} x. \qquad (6.3)$$

Then the following three points can be easily checked:

- $\Gamma^*/S \cong G$.
- S is Noetherian and confluent.
- If w is Britton-reduced and $w \to_S w'$, then w' is Britton-reduced as well.

Now assume that $u = v$ in G. Hence, there exists $w \in \mathsf{IRR}(S)$ such that $u \to_S^* w$ and $v \to_S^* w$. But since u and v are Britton-reduced, only rules of the form (6.3) can be applied in the derivations $u \to_S^* w$ and $v \to_S^* w$. These derivations can be seen as van Kampen diagrams as in Fig. 6.1 for uw^{-1} and wv^{-1}, respectively. By gluing these diagrams together along the w-path, we obtain a van Kampen diagram as in Fig. 6.1. $\qquad\qquad\square$

The semi-Thue system in (6.2) and (6.3) can be used to give a polynomial time Turing-reduction from the word problem for G to the word problem for H in case the groups A_i and B_i are finite.

Example 6.5. Consider the HNN-extension $G = \langle a, t \mid t^{-1}a^2t = a^3 \rangle$. This is the Baumslag-Solitar group $\mathrm{BS}(2,3)$. We have $G \cong \langle \mathbb{Z}, t \mid t^{-1}xt = \varphi(x) \; (x \in 2\mathbb{Z}) \rangle$, where $\varphi : 2\mathbb{Z} \to 3\mathbb{Z}$ is defined by $\varphi(2x) = 3x$. The word $u = ta^2t^{-1}a^3t$ is Britton-reduced, whereas the word $v = ta^3t^{-1}a^2t$ is not Britton-reduced. We have $v = a^4t = ta^6$ in G and $u = ta^5t^{-1}at$ in G.

The following lemma can be used in order to compute a Britton-reduced word for the group element represented by the concatenation of two Britton-reduced words. Recall that for a free group an irreducible word equivalent to the concatenation of two irreducible words u and v can be computed by computing longest words $u_1 = a_1 \cdots a_m$ and $v_0 = a_m^{-1} \cdots a_1^{-1}$ such that $u = u_0 u_1$ and $v = v_0 v_1$. Then $u_0 v_1$ is irreducible. The same idea holds in the HNN-extension G from (6.1).

Lemma 6.6. *Assume that* $u = u_0 t_{i_1}^{\alpha_1} u_1 \cdots t_{i_n}^{\alpha_n} u_n$ *and* $v = v_0 t_{j_1}^{\beta_1} v_1 \cdots t_{j_m}^{\beta_m} v_m$ *are Britton-reduced words. Let* $d(u,v)$ *be the largest number* $d \geq 0$ *such that:*

(a) $A_{i_{n-d+1}}(\alpha_{n-d+1}) = A_{j_d}(-\beta_d)$ *(we set* $A_{i_{n+1}}(\alpha_{n+1}) = A_{j_0}(-\beta_0) = 1$) *and*
(b) *there exists* $c \in A_{j_d}(-\beta_d)$ *with*

$$t_{i_{n-d+1}}^{\alpha_{n-d+1}} u_{n-d+1} \cdots t_{i_n}^{\alpha_n} u_n v_0 t_{j_1}^{\beta_1} \cdots v_{d-1} t_{j_d}^{\beta_d} = c$$

in the group G *from (6.1) (note that this condition is satisfied for* $d = 0$*).*

Moreover, let $c(u,v) \in A_{j_d}(-\beta_d)$ *be the element* c *in (b) (for* $d = d(u,v)$*). Then*

$$u_0 t_{i_1}^{\alpha_1} u_1 \cdots t_{i_{n-d(u,v)}}^{\alpha_{n-d(u,v)}} \left(u_{n-d(u,v)} \, c(u,v) \, v_{d(u,v)} \right) t_{j_{d(u,v)+1}}^{\beta_{d(u,v)+1}} v_{d(u,v)+1} \cdots t_{j_m}^{\beta_m} v_m$$

is a Britton-reduced word equal to uv *in* G*.*

Lemma 6.6 is visualized by Fig. 6.2.

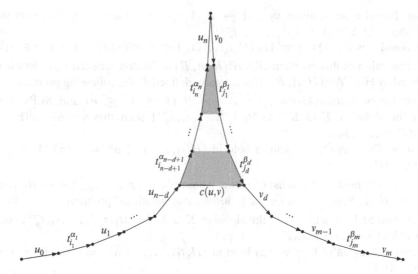

Fig. 6.2 Lemma 6.6

6.2 The Main Computational Problems

The main goal of this chapter is to show that the compressed word problem for an HNN-extension of the form (6.1) is polynomial time Turing-reducible to the compressed word problem for H. In fact, we will prove the existence of such a reduction for a slightly more general problem, which we introduce below.

For further consideration, let us fix the finitely generated group H together with the finite subgroups A and B. Let Σ be a finite generating set for H. These data are fixed, i.e., they will not belong to the input of computational problems.

In the following, when writing down a multiple HNN-extension

$$\langle H, t_1, \ldots, t_n \mid a^{t_i} = \varphi_i(a) \ (1 \leq i \leq n, a \in A) \rangle, \tag{6.4}$$

we assume implicitly that every φ_i is in fact an isomorphism between subgroups $A_1 \leq A$ and $B_1 \leq B$. Hence, φ_i can be viewed as a *partial* isomorphism from our fixed subgroup A to our fixed subgroup B, and (6.4) is in fact an abbreviation for the group

$$\langle H, t_1, \ldots, t_n \mid a^{t_i} = \varphi_i(a) \ (1 \leq i \leq n, a \in \operatorname{dom}(\varphi_i)) \rangle.$$

Note that there is only a fixed number of partial isomorphisms from A to B, but we allow $\varphi_i = \varphi_j$ for $i \neq j$ in (6.4).

Let us introduce several restrictions and extensions of $\mathsf{CWP}(G)$. Our most general problem is the following computational problem $\mathsf{UCWP}(H, A, B)$ (the letter "U" stands for "uniform," meaning that a list of partial isomorphisms from A to B is part of the input):

input: Partial isomorphisms $\varphi_i : A \to B$ $(1 \leq i \leq n)$ and an SLP $\mathbb{A}$ over the alphabet $\Sigma \cup \Sigma^{-1} \cup \{t_1, t_1^{-1}, \ldots, t_n, t_n^{-1}\}$.

question: Does $\mathsf{val}(\mathbb{A}) = 1$ hold in $\langle H, t_1, \ldots, t_n \mid a^{t_i} = \varphi_i(a) \ (1 \leq i \leq n, a \in A)\rangle$?

The restriction of this problem $\mathsf{UCWP}(H, A, B)$ to Britton-reduced input words is denoted by $\mathsf{RUCWP}(H, A, B)$. It is formally defined as the following problem:

input: Partial isomorphisms $\varphi_i : A \to B$ $(1 \leq i \leq n)$ and SLPs $\mathbb{A}, \mathbb{B}$ over the alphabet $\Sigma \cup \Sigma^{-1} \cup \{t_1, t_1^{-1}, \ldots, t_n, t_n^{-1}\}$ such that $\mathsf{val}(\mathbb{A}), \mathsf{val}(\mathbb{B}) \in \mathrm{Red}(H, \varphi_1, \ldots, \varphi_n)$.

question: Does $\mathsf{val}(\mathbb{A}) = \mathsf{val}(\mathbb{B})$ hold in $\langle H, t_1, \ldots, t_n \mid a^{t_i} = \varphi_i(a) \ (1 \leq i \leq n, a \in A)\rangle$?

Let us now consider a fixed list of partial isomorphisms $\varphi_1, \ldots, \varphi_n : A \to B$. Then $\mathsf{RCWP}(H, A, B, \varphi_1, \ldots, \varphi_n)$ is the following computational problem:

input: Two SLPs $\mathbb{A}$ and $\mathbb{B}$ over the alphabet $\Sigma \cup \Sigma^{-1} \cup \{t_1, t_1^{-1}, \ldots, t_n, t_n^{-1}\}$ such that $\mathsf{val}(\mathbb{A}), \mathsf{val}(\mathbb{B}) \in \mathrm{Red}(H, \varphi_1, \ldots, \varphi_n)$.

question: Does $\mathsf{val}(\mathbb{A}) = \mathsf{val}(\mathbb{B})$ hold in $\langle H, t_1, \ldots, t_n \mid a^{t_i} = \varphi_i(a) \ (1 \leq i \leq n, a \in A)\rangle$?

The main result of this chapter is

Theorem 6.7. $\mathsf{UCWP}(H, A, B) \leq_T^P \mathsf{CWP}(H)$.

The rest of Chap. 6 is concerned with the proof of Theorem 6.7.

Let us first consider the special case of the compressed word problem for an HNN-extension of the form (6.1) with $H = A_1 = \cdots = A_n = B_1 = \cdots = B_n$ finite. In this case, we can even assume that the finite group H (represented by its multiplication table) is part of the input.

Lemma 6.8. *The following problem can be solved in polynomial time:*

input: A finite group H, automorphisms $\varphi_i : H \to H$ $(1 \leq i \leq n)$, and an SLP $\mathbb{A}$ over the alphabet $H \cup \{t_1, t_1^{-1}, \ldots t_n, t_n^{-1}\}$.

question: Does $\mathsf{val}(\mathbb{A}) = 1$ hold in $\langle H, t_1, \ldots, t_n \mid h^{t_i} = \varphi_i(h) \ (1 \leq i \leq n, h \in H)\rangle$?

Proof. Let $s \in (H \cup \{t_1, t_1^{-1}, \ldots t_n, t_n^{-1}\})^*$. From the relators of the group $G = \langle H, t_1, \ldots, t_n \mid h^{t_i} = \varphi_i(h) \ (1 \leq i \leq n, h \in H)\rangle$, it follows that there exists a unique $h \in H$ with $s = \pi_t(s)h$ in G. Hence, $s = 1$ in G if and only if $\pi_t(s) = 1$ in the free group $F(t_1, \ldots, t_n)$ and $h = 1$ in H.

Now, let $\mathbb{A}$ be an SLP over the alphabet $H \cup \{t_1, t_1^{-1}, \ldots t_n, t_n^{-1}\}$. Without loss of generality assume that $\mathbb{A}$ is in Chomsky normal form. We can produce an SLP for the projection $\pi_t(\mathsf{val}(\mathbb{A}))$ [see Proposition 3.9(4)] and check in polynomial time whether this SLP produces a word that is the identity in the free group $F(t_1, \ldots, t_n)$; see Theorem 4.14. Hence, it suffices to compute for every variable A of $\mathbb{A}$ the unique $h_A \in H$ with $\mathsf{val}(A) = \pi_t(\mathsf{val}(A))h_A$ in G. We compute the elements h_A bottom-up. The case that the right-hand side for A is a terminal symbol from $H \cup \{t_1, t_1^{-1}, \ldots t_n, t_n^{-1}\}$ is clear. Hence, assume that $\mathrm{rhs}(A) = BC$ and assume that $h_B, h_C \in H$ are already computed. In G we have

$$\mathsf{val}(A) = \mathsf{val}(B)\mathsf{val}(C) = \pi_t(\mathsf{val}(B))h_B\pi_t(\mathsf{val}(C))h_C.$$

Thus, it suffices to compute the unique $h \in H$ with $h_B \pi_t(\mathsf{val}(C)) = \pi_t(\mathsf{val}(C)) h$ in G. Note that if $\pi_t(\mathsf{val}(C)) = t_{i_1}^{\alpha_1} t_{i_2}^{\alpha_2} \cdots t_{i_n}^{\alpha_n}$, then

$$h = \varphi_{i_n}^{\alpha_n}(\cdots \varphi_{i_2}^{\alpha_2}(\varphi_{i_1}^{\alpha_1}(h_B)) \cdots) = (\varphi_{i_1}^{\alpha_1} \cdots \varphi_{i_n}^{\alpha_n})(h_B).$$

The automorphism $f = \varphi_{i_1}^{\alpha_1} \cdots \varphi_{i_n}^{\alpha_n}$ can be easily computed from an SLP $\mathbb{C}$ for the word $\pi_t(\mathsf{val}(C))$ by replacing in $\mathbb{C}$ the terminal symbol t_i (respectively, t_i^{-1}) by φ_i (respectively, φ_i^{-1}). This allows to compute f bottom-up and then to compute $f(h_B)$.   □

Note that the group $\langle H, t_1, \ldots, t_n \mid h^{t_i} = \varphi_i(h) \ (1 \le i \le n, h \in H) \rangle$ is the semidirect product $H \rtimes_\varphi F$, where $F = F(t_1, \ldots, t_n)$ is the free group generated by $t_1, \ldots, t_n$ and the homomorphism $\varphi : F \to \mathrm{Aut}(H)$ is defined by $\varphi(t_i) = \varphi_i$.

6.3 Reducing to Britton-Reduced Sequences

As a first step in the proof of Theorem 6.7, we show that the problem UCWP(H, A, B) is polynomial time Turing-reducible to the problem RUCWP (H, A, B), where all input words are assumed to be Britton-reduced. Later, it will be important that the Turing-reduction from UCWP(H, A, B) to RUCWP(H, A, B) does not change the list of partial isomorphisms $\varphi_1, \ldots, \varphi_n : A \to B$.

Lemma 6.9. *We have* UCWP$(H, A, B) \le_T^P$ RUCWP(H, A, B). *More precisely, there is a polynomial time Turing-reduction from the problem* UCWP(H, A, B) *to the problem* RUCWP(H, A, B) *that on input* $(\varphi_1, \ldots, \varphi_n, \mathbb{A})$ *only asks* RUCWP(H, A, B)-*queries of the form* $(\varphi_1, \ldots, \varphi_n, \mathbb{A}', \mathbb{B}')$ *(thus, the list of partial isomorphisms is not changed).*

Proof. Consider partial isomorphisms $\varphi_i : A \to B$ $(1 \le i \le n)$ and let

$$G = \langle H, t_1, \ldots, t_n \mid a^{t_i} = \varphi_i(a) \ (1 \le i \le n, a \in A) \rangle.$$

Moreover, let $\mathbb{A}$ be an SLP in Chomsky normal form over the alphabet $\Sigma \cup \Sigma^{-1} \cup \{t_1, t_1^{-1}, \ldots, t_n, t_n^{-1}\}$. Using oracle access to RUCWP(H, A, B), we will construct a CSLP $\mathbb{A}'$ with $\mathsf{val}(\mathbb{A}') = \mathsf{val}(\mathbb{A})$ in G and $\mathsf{val}(\mathbb{A}')$ Britton-reduced, on which finally the RUCWP(H, A, B)-oracle can be asked whether $\mathsf{val}(\mathbb{A}') = 1$ in G. The system $\mathbb{A}'$ has the same variables as $\mathbb{A}$ but for every variable X, $\mathsf{val}_{\mathbb{A}'}(X)$ is Britton-reduced and $\mathsf{val}(\mathbb{A}', X) = \mathsf{val}_{\mathbb{A}}(X)$ in G.

Assume that X is a variable of $\mathbb{A}$ with $\mathrm{rhs}(X) = YZ$, where Y and Z were already processed during our reduction process. Hence, $\mathsf{val}(Y)$ and $\mathsf{val}(Z)$ are Britton-reduced. Let

$$\mathsf{val}(Y) = u_0 t_{i_1}^{\alpha_1} u_1 \cdots t_{i_\ell}^{\alpha_\ell} u_\ell \quad \text{and} \quad \mathsf{val}(Z) = v_0 t_{j_1}^{\beta_1} v_1 \cdots t_{j_m}^{\beta_m} v_m.$$

with $u_0, \ldots, u_\ell, v_0, \ldots, v_m \in (\Sigma \cup \Sigma^{-1})^*$ and $\alpha_1, \ldots, \alpha_\ell, \beta_1, \ldots, \beta_k \in \{-1, 1\}$. For $1 \le k \le \ell$ (respectively, $1 \le k \le m$) let $p(k)$ (respectively, $q(k)$) be the unique position within $\mathsf{val}(Y)$ (respectively, $\mathsf{val}(Z)$) such that $\mathsf{val}(Y)[:p(k)] = u_0 t_{i_1}^{\alpha_1} u_1 \cdots t_{i_k}^{\alpha_k}$ (respectively, $\mathsf{val}(Z)[:q(k)] = v_0 t_{j_1}^{\beta_1} v_1 \cdots t_{j_k}^{\beta_k}$). By Proposition 3.9(6) the positions $p(k)$ and $q(k)$ can be computed in polynomial time from k.

According to Lemma 6.6 it suffices to find $d = d(\mathsf{val}(Y), \mathsf{val}(Z)) \in \mathbb{N}$ and $c = c(\mathsf{val}(Y), \mathsf{val}(Z)) \in A \cup B$ in polynomial time. This can be done using binary search: First, compute $\min\{l, m\}$. For a given number $k \le \min\{\ell, m\}$ we want to check whether

$$t_{i_{\ell-k+1}}^{\alpha_{\ell-k+1}} u_{\ell-k+1} \cdots t_{i_\ell}^{\alpha_\ell} u_\ell v_0 t_{j_1}^{\beta_1} \cdots v_{k-1} t_{j_k}^{\beta_k} \in A_{i_{\ell-k+1}}(\alpha_{\ell-k+1}) = A_{j_k}(-\beta_k) \quad (6.5)$$

in the group G. Note that (6.5) is equivalent to $t_{i_{\ell-k+1}}^{\alpha_{\ell-k+1}} = t_{j_k}^{-\beta_k}$ and

$$\exists c \in A_{j_k}(-\beta_k) : (\mathsf{val}(Y)[p(\ell - k + 1) :])^{-1} c = \mathsf{val}(Z)[:q(k)]. \quad (6.6)$$

The two sides of this equation are Britton-reduced words, and the number of possible values $c \in A_{j_k}(-\beta_k)$ is bounded by a constant. Hence, (6.6) is equivalent to a constant number of $\mathsf{RUCWP}(H, A, B)$-instances that can be computed in polynomial time.

In order to find with binary search the value d (i.e., the largest $k \ge 0$ such that (6.5) holds), one has to observe that (6.5) implies that (6.5) also holds for every smaller value k; this follows from Lemma 6.4. From d, we can compute in polynomial time the positions $p(\ell - d + 1)$ and $q(d)$. Then, according to Lemma 6.6, the word

$$\mathsf{val}(Y)[:p(\ell - d + 1) - 1] c \, \mathsf{val}(Z)[q(d) + 1 :]$$

is Britton-reduced and equal to $\mathsf{val}(Y)\mathsf{val}(Z)$ in G. Hence, we can set

$$\mathsf{rhs}(X) := Y[:p(\ell - d + 1) - 1] c \, Z[q(d) + 1 :].$$

This concludes the proof of the lemma.   □

The above proof can be also used in order to derive the following statement:

Lemma 6.10. *Let $\varphi_1, \ldots, \varphi_n : A \to B$ be fixed partial isomorphisms. Then, the problem $\mathsf{CWP}(\langle H, t_1, \ldots, t_n \mid a^{t_i} = \varphi_i(a) \, (1 \le i \le n, a \in A)\rangle)$ is polynomial time Turing-reducible to the problem $\mathsf{RCWP}(H, A, B, \varphi_1, \ldots, \varphi_n)$.*

6.4 Reduction to a Constant Number of Stable Letters

In this section, we show that the number of different stable letters can be reduced to a constant. For this, it is important to note that the associated subgroups $A, B \le H$ do not belong to the input; so their size is a fixed constant.

Fix the constant $\delta = 2 \cdot |A|! \cdot 2^{|A|}$ for the rest of Chap. 6. Note that the number of HNN-extensions of the form $\langle H, t_1, \ldots, t_k \mid a^{t_i} = \psi_i(a)\ (1 \leq i \leq k, a \in A)\rangle$ with $k \leq \delta$ is a constant in our consideration. The following lemma says that RUCWP(H, A, B) can be reduced in polynomial time to one of the problems RCWP$(H, A, B, \psi_1, \ldots, \psi_k)$. Moreover, we can determine in polynomial time which of these problems arises.

Lemma 6.11. *There exists a polynomial time algorithm for the following:*

input: Partial isomorphisms $\varphi_1, \ldots, \varphi_n : A \to B$ and SLPs $\mathbb{A}, \mathbb{B}$ over the alphabet $\Sigma \cup \Sigma^{-1} \cup \{t_1, t_1^{-1}, \ldots t_n, t_n^{-1}\}$ such that $\mathsf{val}(\mathbb{A}), \mathsf{val}(\mathbb{B}) \in \mathrm{Red}(H, \varphi_1, \ldots, \varphi_n)$.
output: Partial isomorphisms $\psi_1, \ldots, \psi_k : A \to B$ where $k \leq \delta$ and SLPs $\mathbb{A}', \mathbb{B}'$ over the alphabet $\Sigma \cup \Sigma^{-1} \cup \{t_1, t_1^{-1}, \ldots t_k, t_k^{-1}\}$ such that:

- *For every $1 \leq i \leq k$ there exists $1 \leq j \leq n$ with $\psi_i = \varphi_j$.*
- $\mathsf{val}(\mathbb{A}'), \mathsf{val}(\mathbb{B}') \in \mathrm{Red}(H, \psi_1, \ldots, \psi_k)$
- $\mathsf{val}(\mathbb{A}) = \mathsf{val}(\mathbb{B})$ *in* $\langle H, t_1, \ldots, t_n \mid a^{t_i} = \varphi_i(a)\ (1 \leq i \leq n, a \in A)\rangle$ *if and only if* $\mathsf{val}(\mathbb{A}') = \mathsf{val}(\mathbb{B}')$ *in* $\langle H, t_1, \ldots, t_k \mid a^{t_i} = \psi_i(a)\ (1 \leq i \leq k, a \in A)\rangle$.

Proof. Fix an instance $(\varphi_1, \ldots, \varphi_n, \mathbb{A}, \mathbb{B})$ of the problem RUCWP(H, A, B). In particular, $\mathsf{val}(\mathbb{A}), \mathsf{val}(\mathbb{B}) \in \mathrm{Red}(H, \varphi_1, \ldots, \varphi_n)$. Define the function $\tau : \{1, \ldots, n\} \to \{1, \ldots, n\}$ by

$$\tau(i) = \min\{k \mid \varphi_k = \varphi_i\}.$$

This mapping can be easily computed in polynomial time from the sequence $\varphi_1, \ldots, \varphi_n$. Assume without loss of generality that the range of τ is $\{1, \ldots, \gamma\}$ for some $\gamma \leq n$. Note that $\gamma \leq |A|! \cdot 2^{|A|} = \frac{\delta}{2}$. For every t_i $(1 \leq i \leq \gamma)$ we take two stable letters $t_{i,0}$ and $t_{i,1}$. Hence, the total number of stable letters is at most δ. We define the HNN-extension

$$G' = \langle H, t_{1,0}, t_{1,1}, \ldots, t_{\gamma,0}, t_{\gamma,1} \mid a^{t_{i,k}} = \varphi_i(a)\ (1 \leq i \leq \gamma, k \in \{0, 1\}, a \in A)\rangle.$$

This HNN-extension has $2\gamma \leq \delta$ many stable letters; it is the HNN-extension $\langle H, t_1, \ldots, t_k \mid a^{t_i} = \psi_i(a)\ (1 \leq i \leq k, a \in A)\rangle$ from the lemma.

It is straightforward to construct a deterministic rational transducer $\mathcal{T}$, which transforms the input word $u_0 t_{i_1}^{\alpha_1} u_1 t_{i_2}^{\alpha_2} u_2 t_{i_3}^{\alpha_3} u_3 \cdots t_{i_m}^{\alpha_m} u_m$ (with $u_0, \ldots, u_m \in (\Sigma \cup \Sigma^{-1})^*$ and $1 \leq i_1, \ldots, i_m \leq n$) into the word

$$[\![\mathcal{T}]\!](w) = u_0 t_{\tau(i_1),1}^{\alpha_1} u_1 t_{\tau(i_2),0}^{\alpha_2} u_2 t_{\tau(i_3),1}^{\alpha_3} u_3 \cdots t_{\tau(i_m),m\bmod 2}^{\alpha_m} u_m.$$

Claim. Let $u, v \in \mathrm{Red}(H, \varphi_1, \ldots, \varphi_n)$ be Britton-reduced. Then $[\![\mathcal{T}]\!](u)$ and $[\![\mathcal{T}]\!](v)$ are also Britton-reduced. Moreover, the following are equivalent:

(a) $u = v$ in $\langle H, t_1, \ldots, t_n \mid a^{t_i} = \varphi_i(a)\ (1 \leq i \leq n, a \in A)\rangle$
(b) $[\![\mathcal{T}]\!](u) = [\![\mathcal{T}]\!](v)$ in the HNN-extension G' and $\pi_t(u) = \pi_t(v)$.

Proof of the claim. Let $u = u_0 t_{i_1}^{\alpha_1} u_1 \cdots t_{i_\ell}^{\alpha_\ell} u_\ell$ and $v = v_0 t_{j_1}^{\beta_1} v_1 \cdots t_{j_m}^{\beta_m} v_m$. The first statement is obvious due to the fact that $[\![\mathcal{T}]\!](u)$ does not contain a subword of the form $t_{i,k}^\alpha w t_{j,k}^\beta$ for $k \in \{0,1\}$, and similarly for $[\![\mathcal{T}]\!](v)$.

For $(a) \Rightarrow (b)$ assume that $u = v$ in $\langle H, t_1, \ldots, t_n \mid a^{t_i} = \varphi_i(a) \ (1 \leq i \leq n, a \in A)\rangle$. Lemma 6.4 implies that $\pi_t(u) = \pi_t(v)$ (i.e., $\ell = m$, $\alpha_1 = \beta_1, \ldots, \alpha_m = \beta_m$, $i_1 = j_1, \ldots, i_m = j_m$) and that there exists a van Kampen diagram of the following form:

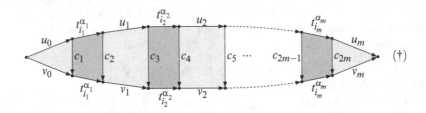

(†)

The relators of G' imply that the following is a valid van Kampen diagram in G':

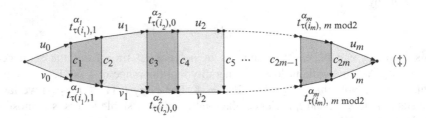

(‡)

Hence, $[\![\mathcal{T}]\!](u) = [\![\mathcal{T}]\!](v)$ in G'.

For $(b) \Rightarrow (a)$ assume that $[\![\mathcal{T}]\!](u) = [\![\mathcal{T}]\!](v)$ in G' and $\pi_t(u) = \pi_t(v)$. We have already argued that $[\![\mathcal{T}]\!](u)$ and $[\![\mathcal{T}]\!](v)$ are Britton-reduced. Hence, $[\![\mathcal{T}]\!](u) = [\![\mathcal{T}]\!](v)$ in G' implies that there exists a van Kampen diagram of the form (‡). Since $\pi_t(u) = \pi_t(v)$ we can obtain a diagram of the form (†) by replacing the dark-shaded t-faces in (‡) by the corresponding t-faces of G. This proofs the claim. □

By the previous claim, $[\![\mathcal{T}]\!](\text{val}(\mathbb{A}))$ and $[\![\mathcal{T}]\!](\text{val}(\mathbb{B}))$ are Britton-reduced. Moreover, SLPs $\mathbb{A}'$ and $\mathbb{B}'$ for these words can be computed in polynomial time by Theorem 3.10. In case $\pi_t(\text{val}(\mathbb{A})) \neq \pi_t(\text{val}(\mathbb{B}))$ (this can be checked in polynomial time by Proposition 3.9(4) and Theorem 3.17), we choose these SLPs such that, e.g., $\text{val}(\mathbb{A}') = t_1$ and $\text{val}(\mathbb{B}') = t_1^{-1}$. Hence, $\text{val}(\mathbb{A}') = \text{val}(\mathbb{B}')$ in G' if and only if $\text{val}(\mathbb{A}) = \text{val}(\mathbb{B})$ in $\langle H, t_1, \ldots, t_n \mid a^{t_i} = \varphi_i(a)(1 \leq i \leq n, a \in A)\rangle$. This proves the lemma. □

Due to Lemma 6.11 it suffices to concentrate our effort on problems of the form $\text{RCWP}(H, A, B, \varphi_1, \ldots, \varphi_k)$, where $k \leq \delta$. Let

$$G_0 = \langle H, t_1, \ldots, t_k \mid a^{t_i} = \varphi_i(a) \ (1 \leq i \leq k, a \in A)\rangle \qquad (6.7)$$

and let us choose $i \in \{1, \ldots, k\}$ such that $|\text{dom}(\varphi_i)|$ is maximal. Without loss of generality assume that $i = 1$. Let $\text{dom}(\varphi_1) = A_1 \leq A$ and let $B_1 = \varphi_1(A_1) \leq B$ be the range of φ_1. We write t for t_1 in the following and define

$$\Gamma = \Sigma \cup \{t_2, \ldots, t_k\}.$$

We can write our HNN-extension G_0 from (6.7) as

$$G_0 = \langle K, t \mid a^t = \varphi_1(a) \ (a \in A_1) \rangle, \tag{6.8}$$

where

$$K = \langle H, t_2, \ldots, t_k \mid a^{t_i} = \varphi_i(a) \ (2 \leq i \leq k, a \in A) \rangle. \tag{6.9}$$

The latter group K is generated by Γ. We may have $k = 1$ in which case $K = H$. The goal of the next three Sects. 6.5–6.7 is to prove the following crucial lemma:

Lemma 6.12. RCWP$(H, A, B, \varphi_1, \ldots, \varphi_k)$ *is polynomial time Turing-reducible to the problems* RCWP$(H, A, B, \varphi_2, \ldots, \varphi_k)$ *and* RUCWP(A_1, A_1, A_1).

6.5 Abstracting from the Base Group K

Our aim in this subsection will be to reduce the compressed word problem for G_0 from (6.8) to the compressed word problem for another group, where we have abstracted from most of the concrete structure of the base group K in (6.9).

Let us consider an input $(\mathbb{A}, \mathbb{B})$ for RCWP$(H, A, B, \varphi_1, \ldots, \varphi_k)$ with $k \leq \delta$. Without loss of generality assume that $k = \delta$. Thus, $\mathbb{A}$ and $\mathbb{B}$ are SLPs over the alphabet $\Sigma \cup \Sigma^{-1} \cup \{t_1, t_1^{-1}, \ldots, t_\delta, t_\delta^{-1}\} = \Gamma \cup \Gamma^{-1} \cup \{t, t^{-1}\}$ with $\text{val}(\mathbb{A}), \text{val}(\mathbb{B}) \in \text{Red}(H, \varphi_1, \ldots, \varphi_\delta)$. Hence, we also have $\text{val}(\mathbb{A}), \text{val}(\mathbb{B}) \in \text{Red}(K, \varphi_1)$.

Without loss of generality we may assume that $\pi_t(\text{val}(\mathbb{A})) = \pi_t(\text{val}(\mathbb{B}))$. This property can be checked in polynomial time using Proposition 3.9(4) and Theorem 3.17. If it is not satisfied, then we have $\text{val}(\mathbb{A}) \neq \text{val}(\mathbb{B})$ in G_0. Hence, there are $m \geq 0$, $\alpha_1, \ldots, \alpha_m \in \{1, -1\}$, and words $u_0, v_0 \ldots, u_m, v_m \in (\Gamma \cup \Gamma^{-1})^*$ such that

$$\text{val}(\mathbb{A}) = u_0 t^{\alpha_1} u_1 \cdots t^{\alpha_m} u_m \text{ and} \tag{6.10}$$

$$\text{val}(\mathbb{B}) = v_0 t^{\alpha_1} v_1 \cdots t^{\alpha_m} v_m. \tag{6.11}$$

One might think that the number of different words u_i (respectively, v_i) may grow exponentially in the size of $\mathbb{A}$ (respectively, $\mathbb{B}$). But we will see that this is actually not the case.

Let us replace every occurrence of t^α ($\alpha \in \{1, -1\}$) in $\mathbb{A}$ and $\mathbb{B}$ by $aa^{-1}t^\alpha aa^{-1}$, where $a \in \Gamma$ is arbitrary. This is to ensure that any two occurrences of symbols from $\{t, t^{-1}\}$ are separated by a nonempty word over $\Gamma \cup \Gamma^{-1}$, i.e., we can assume that $u_0, v_0, \ldots, u_m, v_m \in (\Gamma \cup \Gamma^{-1})^+$ in (6.10) and (6.11).

Our first goal is to transform $\mathbb{A}$ (and similarly $\mathbb{B}$) into an equivalent SLP that generates in a first phase a word of the form $X_0 t^{\alpha_1} X_1 \cdots t^{\alpha_m} X_m$, where X_i is a further variable that generates in a second phase the word $u_i \in (\Gamma \cup \Gamma^{-1})^+$. This is similar to the notion of a 2-level PCSLP from Definition 3.24. Assume that the SLP $\mathbb{A} = (U, \{t, t^{-1}\} \cup \Gamma \cup \Gamma^{-1}, S, \text{rhs})$ is in Chomsky normal form.

In a first step, we remove every variable $X \in U$ from $\mathbb{A}$ such that $\text{rhs}(X) \in \{t, t^{-1}\}$ by replacing X in all right-hand sides of $\mathbb{A}$ by $\text{rhs}(X)$. Now, all right-hand sides of $\mathbb{A}$ are of the form $YZ, t^\alpha Z, Yt^\alpha$, or $x \in \Gamma \cup \Gamma^{-1}$, where $Y, Z \in U$.

Next we split the set U of variables of $\mathbb{A}$ into two parts:

$$U_K = \{X \in U \mid \text{val}(X) \in (\Gamma \cup \Gamma^{-1})^+\} \quad \text{and} \quad U_t = U \setminus U_K^0.$$

The subscript K refers to the fact that every variable from U_K defines an element from the new base group $K \le G_0$, whereas the subscript t refers to the fact that every variable from U_t generates a word where K-generators as well as t or t^{-1} occurs.

Now we manipulate all right-hand sides for variables from U_t in a bottom-up process. Thereby we add further variables to U_K, whereas the set U_t will not change. At each stage, the tuple

$$\mathbb{A}_t := (U_t, \{t, t^{-1}\} \cup U_K, S, \text{rhs} \restriction_{U_t})$$

is a CSLP that generates a word from $(U_K)^+ t^{\alpha_1} (U_K)^+ \cdots t^{\alpha_m} (U_K)^+$.

Consider a variable $X \in U_t$ such that every variable in $u := \text{rhs}(X)$ is already processed, but X is not yet processed. If u is of the form $t^\alpha Z$ or Yt^α, then there is nothing to do. Now assume that $u = YZ$ such that Y and Z are already processed. Consider the last symbol $\omega \in \{t, t^{-1}\} \cup U_K$ of $\text{val}_{\mathbb{A}_t}(Y)$ and the first symbol $\alpha \in \{t, t^{-1}\} \cup U_K$ of $\text{val}_{\mathbb{A}_t}(Z)$ (these symbols can be computed in polynomial time). If either $\omega \in \{t, t^{-1}\}$ or $\alpha \in \{t, t^{-1}\}$, then again there is nothing to do. Otherwise, $\omega, \alpha \in U_K$. We now set $U_K := U_K \cup \{X'\}$, where X' is a fresh variable with $\text{rhs}(X') = \omega\alpha$. Finally, we redefine $\text{rhs}(X) := Y[: \ell - 1]X'Z[2 :]$, where $\ell = |\text{val}_{\mathbb{A}_t}(Y)|$.

When all variables from U_t are processed, we transform the final CSLP $\mathbb{A}_t$ in polynomial time into an equivalent SLP using Theorem 3.14. Let us denote this SLP again by $\mathbb{A}_t$. Moreover, let

$$\mathbb{A}_K := (U_K, \Gamma \cup \Gamma^{-1}, \text{rhs} \restriction_{U_K}).$$

This is an SLP without initial variable. The construction implies that

$$\text{val}(\mathbb{A}_t) = X_0 t^{\alpha_1} X_1 \cdots t^{\alpha_m} X_m \tag{6.12}$$

with $X_0, \ldots, X_m \in U_K$ and $\mathsf{val}_{\mathbb{A}_K}(X_k) = u_i$. Note that the number of different X_i is polynomially bounded, simply because the set U_K was computed in polynomial time. Hence, also the number of different u_i in (6.10) is polynomially bounded.

For the SLP $\mathbb{B}$ the same procedure yields the following data:

- An SLP $\mathbb{B}_t = (V_t, \{t, t^{-1}\} \cup V_K, S, \mathsf{rhs} \!\restriction_{V_t})$ such that $\mathsf{val}(\mathbb{B}_t) = Y_0 t^{\alpha_1} Y_1 \cdots t^{\alpha_m} Y_m$.
- An SLP $\mathbb{B}_K = (V_K, \Gamma \cup \Gamma^{-1}, \mathsf{rhs} \!\restriction_{V_k})$ without initial variable such that $Y_1, \ldots, Y_m \in V_K$ and $\mathsf{val}_{\mathbb{B}_K}(Y_i) = v_i$.

Without loss of generality assume that $U_K \cap V_K = \emptyset$. Let $W_K = U_K \cup V_K$. We assume that we have a single right-hand side mapping rhs with domain $U_t \cup V_t \cup U_K \cup V_K$. Let $\mathbb{C} = (W_K, \Gamma \cup \Gamma^{-1}, \mathsf{rhs} \!\restriction_{W_K})$. In the following, for $Z \in W_K$ we write $\mathsf{val}(Z)$ for $\mathsf{val}_{\mathbb{C}}(Z) \in (\Gamma \cup \Gamma^{-1})^+$.

Let us next consider the free product $F(W_K) * A_1 * B_1$. Recall that A_1 (respectively, B_1) is the domain (respectively, range) of the partial isomorphism φ_1. Clearly, in this free product, A_1 and B_1 have trivial intersection (even if $|A_1 \cap B_1| > 1$ in H). We now define the finite set of relators

$$\mathscr{E} = \{Z_1 c_1 Z_2^{-1} c_2^{-1} \mid Z_1, Z_2 \in W_K, c_1, c_2 \in A_1 \cup B_1, \mathsf{val}(Z_1) c_1 = c_2 \mathsf{val}(Z_2) \text{ in } K\}. \tag{6.13}$$

We can compute the set $\mathscr{E}$ in polynomial time using oracle access to $\mathsf{CWP}(K)$ or, alternatively, by Lemma 6.10 using oracle access to $\mathsf{RCWP}(H, A, B, \varphi_2, \ldots, \varphi_k)$. This is the only step where we need oracle access to $\mathsf{RCWP}(H, A, B, \varphi_2, \ldots, \varphi_k)$ in the proof of Lemma 6.12. Note that the relator $Z_1 c_1 Z_2^{-1} c_2^{-1}$ stands for the relation $Z_1 c_1 = c_2 Z_2$.

Consider the group

$$\begin{aligned} G_1 &= \langle F(W_K) * A_1 * B_1, t \mid \mathscr{E}, t^{-1} a t = \varphi_1(a) \, (a \in A_1)\rangle \\ &= \langle (F(W_K) * A_1 * B_1)/N, t \mid a^t = \varphi_1(a) \, (a \in A_1)\rangle, \end{aligned}$$

where

$$N = \langle \mathscr{E} \rangle^{F(W_K) * A_1 * B_1}$$

is the normal closure of $\mathscr{E}$ in $F(W_K) * A_1 * B_1$. We can define a homomorphism

$$\psi : F(W_K) * A_1 * B_1 \to K$$

by $\psi(Z) = \mathsf{val}(Z)$ for $Z \in W_K$, $\psi(a) = a$ for $a \in A_1$, and $\psi(b) = b$ for $b \in B_1$. Of course, the restrictions of ψ to A_1 as well as B_1 are injective. Moreover, for every relator $Z_1 c_1 Z_2^{-1} c_2^{-1} \in \mathscr{E}$ we have $\psi(Z_1 c_1 Z_2^{-1} c_2^{-1}) = \mathsf{val}(Z_1) c_1 \mathsf{val}(Z_2)^{-1} c_2^{-1} = 1$ in K. Thus, ψ defines a homomorphism

$$\hat{\psi} : (F(W_K) * A_1 * B_1)/N \to K.$$

Moreover, $A_1 \cap N = 1$: If $a \in N \cap A_1$, then $\psi(a) \in \psi(N) = 1$; thus, $a = 1$, since ψ is injective on A_1. Similarly, $B_1 \cap N = 1$. This means that A_1 and B_1 can be naturally embedded in $(F(W_K) * A_1 * B_1)/N$, and $\varphi_1 : A_1 \to B_1$ can be considered as an isomorphism between the images of this embedding in $(F(W_K) * A_1 * B_1)/N$. Therefore, the above group G_1 is an HNN-extension with base group $(F(W_K) * A_1 * B_1)/N \leq G_1$. Moreover, $\hat{\psi} : (F(W_K) * A_1 * B_1)/N \to K$ can be lifted to a homomorphism

$$\hat{\psi} : G_1 \to G_0 = \langle K, t \mid a^t = \varphi_1(a) \ (a \in A_1) \rangle.$$

Here, G_0 is from (6.8).

The idea for the construction of G_1 is to abstract as far as possible from the concrete structure of the original base group K. We only keep those K-relations that are necessary to prove (or disprove) that $\mathsf{val}(\mathbb{A}) = \mathsf{val}(\mathbb{B})$ in the group G_0. These are the relators from $\mathscr{E}$.

Lemma 6.13. *We have* $\mathsf{val}(\mathbb{A}_t), \mathsf{val}(\mathbb{B}_t) \in \mathrm{Red}((F(W_K) * A_1 * B_1)/N, \varphi_1)$.

Proof. Recall that $\mathsf{val}(\mathbb{A}), \mathsf{val}(\mathbb{B}) \in \mathrm{Red}(K, \varphi_1)$. Consider, for instance, a factor $t^{-1} X_i t$ of $\mathsf{val}(\mathbb{A}_t)$ from (6.12). If $X_i = a$ in $(F(W_K) * A_1 * B_1)/N$ for some $a \in A_1$, then after applying $\hat{\psi}$, we have $\mathsf{val}(X_i) = u_i = a$ in K. Hence, $\mathsf{val}(\mathbb{A})$ from (6.10) would not be Britton-reduced. $\qquad\square$

Lemma 6.14. *The following statements are equivalent:*

(a) $\mathsf{val}(\mathbb{A}) = \mathsf{val}(\mathbb{B})$ *in G_0 from (6.8).*
(b) $\mathsf{val}(\mathbb{A}_t) = \mathsf{val}(\mathbb{B}_t)$ *in G_1.*

Proof. For $(b) \Rightarrow (a)$ assume that $\mathsf{val}(\mathbb{A}_t) = \mathsf{val}(\mathbb{B}_t)$ in G_1. We obtain in G_0 $\mathsf{val}(\mathbb{A}) = \hat{\psi}(\mathsf{val}(\mathbb{A}_t)) = \hat{\psi}(\mathsf{val}(\mathbb{B}_t)) = \mathsf{val}(\mathbb{B})$.

For $(a) \Rightarrow (b)$ assume that $\mathsf{val}(\mathbb{A}) = \mathsf{val}(\mathbb{B})$ in the group G_0. Since $\mathsf{val}(\mathbb{A})$ and $\mathsf{val}(\mathbb{B})$ are Britton-reduced and $\pi_t(\mathsf{val}(\mathbb{A})) = \pi_t(\mathsf{val}(\mathbb{B}))$, we obtain a van Kampen diagram of the form

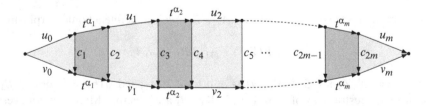

In this diagram, we can replace every light-shaded face, representing the K-relation $u_i c_{2i+1} = c_{2i} v_i$, by a face representing the valid $\mathscr{E}$-relation $X_i c_{2i+1} = c_{2i} Y_i$; see (6.13). We obtain the following van Kampen diagram, which shows that $\mathsf{val}(\mathbb{A}_t) = \mathsf{val}(\mathbb{B}_t)$ in G_1:

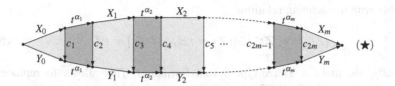

This proves the lemma.   □

By Lemma 6.14, it remains to check whether $\mathsf{val}(\mathbb{A}_t) = \mathsf{val}(\mathbb{B}_t)$ in the HNN-extension G_1, where $\mathsf{val}(\mathbb{A}_t)$ and $\mathsf{val}(\mathbb{B}_t)$ are both Britton-reduced.

6.6   Eliminating Letters from $B_1 \cup \{t, t^{-1}\}$

Recall the Tietze transformations from Sect. 2.3. We can use the defining relations $b = t^{-1}\varphi_1^{-1}(b)t$ ($b \in B_1 \setminus \{1\}$) of G_1 for Tietze transformations of type 4. For this, note that $A_1 \cap B_1 = \{1\}$ in the group $F(W_K) * A_1 * B_1$. In this way we eliminate in the group G_1 the generators from $B_1 \setminus \{1\}$. After this transformation, we may have apart from relations of the form

$$Z_1 a_1 = a_2 Z_2 \text{ with } a_1, a_2 \in A_1 \qquad (6.14)$$

also defining relations of the forms

$$Z_1 t^{-1} a_1 t = a_2 Z_2$$
$$Z_1 a_1 = t^{-1} a_2 t Z_2$$
$$Z_1 t^{-1} a_1 t = t^{-1} a_2 t Z_2,$$

where $a_1, a_2 \in A_1$ (here and in the following, we implicitly assume all valid relations $a_1 a_2 = a$ to get the structure of the free factor A_1). We can replace these relations by relations of the following types

$$Z_1 t^{-1} a_1 = a_2 Z_2 t^{-1} \qquad (6.15)$$
$$t Z_1 a_1 = a_2 t Z_2 \qquad (6.16)$$
$$t Z_1 t^{-1} a_1 = a_2 t Z_2 t^{-1} \qquad (6.17)$$

and end up with the group

$$G_2 = \langle F(W_K) * A_1, t \mid (6.14) - (6.17) \rangle$$

that is isomorphic to G_1. Let us now introduce for every $Z \in W_K$ the new generators

$$[Zt^{-1}], [tZ], [tZt^{-1}]$$

together with the defining relations

$$[Zt^{-1}] = Zt^{-1}, \ [tZ] = tZ, \ [tZt^{-1}] = tZt^{-1}. \tag{6.18}$$

Formally, we make a Tietze-type-3 transformation. This allows to replace the defining relations (6.15)–(6.17) by

$$[Z_1 t^{-1}]a_1 = a_2[Z_2 t^{-1}] \tag{6.19}$$

$$[tZ_1]a_1 = a_2[tZ_2] \tag{6.20}$$

$$[tZ_1 t^{-1}]a_1 = a_2[tZ_2 t^{-1}] \tag{6.21}$$

leading to the group

$$G_3 = \langle F(\{Z, [Zt^{-1}], [tZ], [tZt^{-1}] | Z \in W_K\}) * A_1, t \mid (6.14), (6.18)–(6.21)\rangle. \tag{6.22}$$

Finally, we can eliminate t and t^{-1} by replacing (6.18) by

$$[tZ] = [Zt^{-1}]^{-1}Z^2, \ [tZt^{-1}] = [tZ]Z^{-1}[Zt^{-1}], \tag{6.23}$$

which is a Tietze-type-4 transformation. It yields the group

$$G_4 = \langle F(\{Z, [Zt^{-1}], [tZ], [tZt^{-1}] \mid Z \in W_K\}) * A_1 \mid (6.14), (6.19)–(6.21), (6.23)\rangle.$$

Since each transformation from G_1 to G_4 is a Tietze transformation, G_1 is isomorphic to G_4. We now want to rewrite the SLPs $\mathbb{A}_t$ and $\mathbb{B}_t$ into new SLPs over the generators of G_4. This can be done with a deterministic rational transducer $\mathscr{T}$ that reads a word $X_0 t^{\alpha_1} X_1 t^{\alpha_2} X_2 \cdots t^{\alpha_m} X_m$ \$ from the input tape, where \$ is an end marker and

- replaces every occurrence of a factor tX_i with $\alpha_{i+1} \neq -1$ by the symbol $[tX_i]$,
- replaces every occurrence of a factor $X_i t^{-1}$ with $\alpha_i \neq 1$ by the symbol $[X_i t^{-1}]$, and finally
- replaces every occurrence of a factor $tX_i t^{-1}$ by the symbol $[tX_i t^{-1}]$.

The set of states of the transducer $\mathscr{T}$ is $\{\varepsilon, t, q_f\} \cup \{Z, tZ \mid Z \in W_K\}$, and the transitions are shown in Fig. 6.3 (for all $Z, Z' \in W_k$). The initial state is ε, and the unique final state is q_f. By Theorem 3.10, we can construct in polynomial time SLPs that generate the words $[\![\mathscr{T}]\!](\mathsf{val}(\mathbb{A}_t)\$)$ and $[\![\mathscr{T}]\!](\mathsf{val}(\mathbb{B}_t)\$)$.

Let G_5 be the group that is obtained by removing the relations (6.23) in the above presentation of the group G_4, i.e.,

$$G_5 = \langle F(\{Z, [Zt^{-1}], [tZ], [tZt^{-1}] \mid Z \in W_K\}) * A_1 \mid (6.14), (6.19)–(6.21)\rangle. \tag{6.24}$$

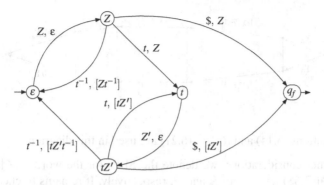

Fig. 6.3 The transitions of the deterministic rational transducer $\mathcal{T}$.

Lemma 6.15. *The following statements are equivalent:*

(a) $\mathsf{val}(\mathbb{A}) = \mathsf{val}(\mathbb{B})$ *in* G_0
(b) $\mathsf{val}(\mathbb{A}_t) = \mathsf{val}(\mathbb{B}_t)$ *in* G_1
(c) $[\![\mathcal{T}]\!](\mathsf{val}(\mathbb{A}_t)\$) = [\![\mathcal{T}]\!](\mathsf{val}(\mathbb{B}_t)\$)$ *in* G_4
(d) $[\![\mathcal{T}]\!](\mathsf{val}(\mathbb{A}_t)\$) = [\![\mathcal{T}]\!](\mathsf{val}(\mathbb{B}_t)\$)$ *in* G_5

Proof. The equivalence of (a) and (b) was stated in Lemma 6.14. The equivalence of (b) and (c) is clear since G_1 and G_4 are isomorphic and the transducer $\mathcal{T}$ rewrites a word over the generators of G_1 into a word over the generators of G_4. Moreover, (d) implies (c) because we omit one type of relations, namely (6.23), when going from G_5 to G_4. It remains to prove that (a) implies (d). If $\mathsf{val}(\mathbb{A}) = \mathsf{val}(\mathbb{B})$ in G_0, then, as argued in the proof of Lemma 6.14, we obtain a van Kampen diagram of the form ($\bigstar$) (p. 129) in the group G_1. The boundary of every light-shaded face is labeled with a relation from $\mathscr{E}$. We obtain a van Kampen diagram for $[\![\mathcal{T}]\!](\mathsf{val}(\mathbb{A}_t)\$) = [\![\mathcal{T}]\!](\mathsf{val}(\mathbb{B}_t)\$)$ in G_5 by removing all vertical edges that connect (i) target nodes of t-labeled edges or (ii) source nodes of t^{-1}-labeled edges. The boundary cycle of every remaining face is labeled with a relator of type (6.14) or (6.19)–(6.21), which are the relators of G_5. An example is shown below.   □

Example 6.16. Let us give an example of the transformation from a diagram of the form ($\bigstar$) into a van Kampen diagram for the group G_5. Assume that the diagram in G_1 is

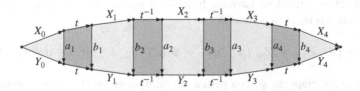

Then we obtain the following van Kampen diagram in the group G_5:

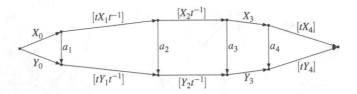

Only the relations (6.14) and (6.19)–(6.21) are used in this diagram.

For the further considerations, we denote the SLPs for the words $[\![\mathscr{T}]\!](\mathsf{val}(\mathbb{A}_t)\$)$ and $[\![\mathscr{T}]\!](\mathsf{val}(\mathbb{B}_t)\$)$ again with $\mathbb{A}$ and $\mathbb{B}$, respectively. It remains to check whether $\mathsf{val}(\mathbb{A}) = \mathsf{val}(\mathbb{B})$ in G_5. Let

$$\mathscr{L} = \{Z, [Zt^{-1}], [tZ], [tZt^{-1}] \mid Z \in W_K\}$$

and let us redefine the set of defining relations $\mathscr{E}$ as the set of all defining relations of the form (6.14), (6.19)–(6.21). Thus,

$$G_5 = \langle F(\mathscr{L}) * A_1 \mid \mathscr{E} \rangle,$$

where every defining relation in $\mathscr{E}$ is of the form $Z_1 a_1 = a_2 Z_2$ for $Z_1, Z_2 \in \mathscr{L}$ and $a_1, a_2 \in A_1$.

6.7   Transforming G_5 into an HNN-Extension

By further Tietze transformations we show that the group $G_5 = \langle F(\mathscr{L}) * A_1 \mid \mathscr{E} \rangle$ is an HNN-extension with the base group A_1 and associated subgroups A_1 and A_1. This will prove Lemma 6.12. To this end, let us take a relation $Z_1 a_1 = a_2 Z_2$ with $Z_1 \neq Z_2$. With a Tietze-type-4 transformation we can eliminate Z_2 by replacing it with $a_2^{-1} Z_1 a_1$. Subwords of the form aa' with $a, a' \in A_1$ that arise after this transformation can of course be multiplied out in the finite group A_1. We carry out the same replacement $Z_2 \mapsto a_2^{-1} Z_1 a_1$ also in the SLPs $\mathbb{A}$ and $\mathbb{B}$, which increases the size of the SLPs only by an additive constant, and repeat these steps. After polynomially many Tietze transformations we obtain a group presentation, where all defining relations have the form $Z = a_1 Z a_2$, i.e., $a_2 = Z^{-1} a_1^{-1} Z$. We can write this group presentation as

$$G_6 = \langle A_1, Z_1, \ldots, Z_m \mid Z_i^{-1} a Z_i = \psi_i(a) \ (1 \le i \le m, a \in \mathsf{dom}(\psi_i)) \rangle.$$

Note that every mapping ψ_i is a partial automorphism on A_1 since it results from the conjugation by some element in our initial group. Hence, G_6 is indeed an HNN-extension over A_1.

We can now finish the proof of Lemma 6.12.

Proof of Lemma 6.12. We have to show that the problem $\mathsf{RCWP}(H, A, B, \varphi_1, \ldots, \varphi_k)$ is polynomial time Turing-reducible to the problems $\mathsf{RCWP}(H, A, B, \varphi_2, \ldots, \varphi_k)$ and $\mathsf{RUCWP}(A_1, A_1, A_1)$. Using oracle access to the problem $\mathsf{RCWP}(H, A, B, \varphi_2, \ldots, \varphi_k)$ [which was necessary for computing the set of defining relations $\mathscr{E}$ from (6.13)], we have computed in polynomial time from a given $\mathsf{RCWP}(H, A, B, \varphi_1, \ldots, \varphi_k)$-instance an $\mathsf{UCWP}(A_1, A_1, A_1)$-instance, which is a positive instance if and only if the original $\mathsf{RCWP}(H, A, B, \varphi_1, \ldots, \varphi_k)$-instance is positive. A final application of Lemma 6.9 allows to reduce $\mathsf{UCWP}(A_1, A_1, A_1)$ to $\mathsf{RUCWP}(A_1, A_1, A_1)$. This proves Lemma 6.12. □

6.8 Final Step in the Proof of Theorem 6.7

We now apply Lemma 6.11 to the problem $\mathsf{RUCWP}(A_1, A_1, A_1)$ (one of the two target problems in Lemma 6.12). An input for this problem can be reduced in polynomial time to an instance of a problem $\mathsf{RCWP}(A_1, A_1, A_1, \psi_1, \ldots, \psi_k)$, where $\psi_1, \ldots, \psi_k : A_1 \to A_1$ and $k \leq 2|A_1|! \cdot 2^{|A_1|} \leq 2|A|! \cdot 2^{|A|} = \delta$. Our current HNN-extension is

$$G_7 = \langle A_1, t_1, \ldots, t_k \mid a^{t_i} = \psi_i(a) \; (1 \leq i \leq k, a \in \mathrm{dom}(\psi_i)) \rangle.$$

We next separate the (constantly many) stable letters $t_1, \ldots, t_k$ that occur in the $\mathsf{RCWP}(A_1, A_1, A_1, \psi_1, \ldots, \psi_k)$-instance into two sets: $\{t_1, \ldots, t_k\} = S_1 \cup S_2$, where $S_1 = \{t_i \mid \mathrm{dom}(\psi_i) = A_1\}$ and $S_2 = \{t_1, \ldots, t_k\} \setminus S_1$. Without loss of generality assume that $S_2 = \{t_1, \ldots, t_\ell\}$. Then we can write the HNN-extension G_7 as

$$G_7 = \langle H', t_1, \ldots, t_\ell \mid a^{t_i} = \psi_i(a) \; (1 \leq i \leq \ell, a \in \mathrm{dom}(\psi_i)) \rangle, \tag{6.25}$$

where

$$H' = \langle A_1, t_{\ell+1}, \ldots, t_k \mid a^{t_i} = \psi_i(a) \; (\ell + 1 \leq i \leq k, a \in A_1) \rangle.$$

Note that (i) $|\mathrm{dom}(\psi_i)| < |A_1|$ for every $1 \leq i \leq \ell$ and that (ii) $A_1 = \mathrm{dom}(\psi_i)$ for every $\ell + 1 \leq i \leq k$. By Lemma 6.8, $\mathsf{CWP}(H')$ can be solved in polynomial time. The group H' is in fact the semidirect product $A_1 \rtimes_\varphi F(t_{\ell+1}, \ldots, t_k)$, where the homomorphism $\varphi : F(t_{\ell+1}, \ldots, t_k) \to \mathrm{Aut}(A_1)$ is defined by $\varphi(t_i) = \psi_i$. Recall also that at the end of Sect. 6.4, A_1 was chosen to be of maximal cardinality among the domains of all partial isomorphisms $\varphi_1, \ldots, \varphi_k$. The following proposition summarizes what we have shown so far:

Proposition 6.17. *Let $\varphi_1, \ldots, \varphi_k : A \to B$ be partial isomorphisms, where $k \leq \delta$, $A_1 = \mathrm{dom}(\varphi_1)$, and, without loss of generality $|A_1| \geq |\mathrm{dom}(\varphi_i)|$ for $1 \leq i \leq k$. The problem $\mathsf{RCWP}(H, A, B, \varphi_1, \ldots, \varphi_k)$ is polynomial time Turing-reducible to the following (constantly many) decision problems:*

(1) $\mathsf{RCWP}(H, A, B, \varphi_2, \ldots, \varphi_k)$,

(2) $\mathsf{RCWP}(A_1 \rtimes_\varphi F, A_1, A_1, \psi_1, \ldots, \psi_\ell)$, where F is a free group of rank at most δ, $\varphi : F \to \mathrm{Aut}(A_1)$ is a homomorphism, $\ell \leq \delta$, and $\psi_1, \ldots, \psi_\ell : A_1 \to A_1$ are partial automorphisms with $|\mathrm{dom}(\psi_i)| < |A_1|$ for all $1 \leq i \leq \ell$.

Note that indeed, in (2) there are only constantly many semidirect products of the form $A_1 \rtimes_\varphi F$ and that $\mathsf{CWP}(A_1 \rtimes_\varphi F)$ can be solved in polynomial time by Lemma 6.8.

Now, we have all the tools needed to prove the main theorem of this chapter:

Proof of Theorem 6.7. By Lemmas 6.9 and 6.11 it suffices to solve a problem $\mathsf{RCWP}(H, A, B, \varphi_1, \ldots, \varphi_k)$ (with $k \leq \delta$) in polynomial time. For this we apply Proposition 6.17 repeatedly, i.e., we apply Proposition 6.17 to the two target problems (1) and (2) from the proposition again. Note that in the target problems one of the two properties holds:

- The maximal size of an associated subgroup is smaller than in the input instance.
- The maximal size of an associated subgroup is the same as in the input instance, but the number of stable letters is smaller than in the input instance.

This implies that after $|A| \cdot \delta = 2 \cdot |A| \cdot |A|! \cdot 2^{|A|}$ (which is a fixed constant) many applications of Proposition 6.17, we have reduced $\mathsf{RCWP}(H, A, B, \varphi_1, \ldots, \varphi_k)$ to the problems $\mathsf{CWP}(H)$ and $\mathsf{CWP}(C \rtimes_\varphi F)$, where $C \leq A$, F is a free group of rank at most δ, and $\varphi : F \to \mathrm{Aut}(C)$ is a homomorphism. Note that this is still a polynomial time Turing-reduction, since the composition of a constant number of polynomial time Turing-reductions is again a polynomial time Turing-reduction. □

6.9 Amalgamated Products

In this section we prove a transfer theorem for the compressed word problem for an amalgamated free product, where the amalgamated subgroups are finite. We will deduce this result from our transfer theorem for HNN-extensions. Let us start with the definition of an amalgamated free product:

Definition 6.18 (amalgamated free product). Let H_1 and H_2 be two groups with subgroups $A_1 \leq H_1$ and $A_2 \leq H_2$. Let $\varphi : A_1 \mapsto A_2$ be an isomorphism. The *amalgamated free product of H_1 and H_2, amalgamating the subgroups A_1 and A_2 by the isomorphism φ,* is the group $G = \langle H_1 * H_2 \mid a = \varphi(a) \ (a \in A_1) \rangle$.

Theorem 6.19. *Let $G = \langle H_1 * H_2 \mid a = \varphi(a)\ (a \in A_1)\rangle$ be an amalgamated free product of finitely generated groups G_1 and G_2 with $A_1 \leq G_1$ a finite subgroup. Then $\mathsf{CWP}(G) \leq^P_T \{\mathsf{CWP}(H_1), \mathsf{CWP}(H_2)\}$.*

Proof. It is well known [119, Theorem 2.6, p. 187] that G can be embedded into the HNN-extension

$$G' := \langle H_1 * H_2, t \mid a^t = \varphi(a)\ (a \in A_1)\rangle$$

by the homomorphism Φ with

$$\Phi(x) = \begin{cases} t^{-1}xt & \text{if } x \in H_1 \\ x & \text{if } x \in H_2. \end{cases}$$

Given an SLP $\mathbb{A}$ over generators of G, we can compute in polynomial time an SLP $\mathbb{B}$ with $\mathsf{val}(\mathbb{B}) = \Phi(\mathsf{val}(\mathbb{A}))$; see Proposition 3.9(4). We obtain

$$\mathsf{val}(\mathbb{A}) = 1 \text{ in } G \iff \Phi(\mathsf{val}(\mathbb{A})) = 1 \text{ in } \Phi(G)$$

$$\iff \mathsf{val}(\mathbb{B}) = 1 \text{ in } G'.$$

By Theorem 6.7 and Corollary 5.5, $\mathsf{CWP}(G')$ is polynomial time Turing-reducible to the problems $\mathsf{CWP}(H_1)$ and $\mathsf{CWP}(H_2)$. $\qquad\square$

Chapter 7
Outlook

We conclude this book with a few remarks about topics related to the compressed word problem for groups, which could not be covered for space reasons.

In this book, we focused on the compressed word problem for groups. But it makes perfect sense to study the compressed word problem for finitely generated monoids as well. If M is a finitely generated monoid with a finite generating set Σ, then the compressed word problem for M asks whether, for given SLPs $\mathbb{A}$ and $\mathbb{B}$ over Σ, $\mathsf{val}(\mathbb{A}) = \mathsf{val}(\mathbb{B})$ holds in M. As for groups, the complexity of this problem is independent of the concrete generating set.

Implicitly, we have seen some complexity statements for compressed word problems for monoids. Theorem 3.17 states that the compressed word problem for a free monoid can be solved in polynomial time, and Lemma 5.24 generalizes this statement to trace monoids. In [113] it was shown that the compressed word problem for any finitely generated free inverse monoid of rank at least two is complete for the level Π_2^p of the polynomial time hierarchy. Inverse monoids are an important class of monoids that have inverses in a weaker sense than groups have; see [103]. The polynomial time hierarchy is a hierarchy of complexity classes between **P** and **PSPACE** that is believed to be proper. In particular, the class Π_2^p contains **NP** and **coNP**, and it is conjectured that these inclusions are proper. The ordinary word problem for a free inverse monoid can be solved in logarithmic space [115].

We have seen that the compressed word problem for a group G can be used in order to solve the word problem for the automorphism group $\mathsf{Aut}(G)$ (or a finitely generated subgroup of $\mathsf{Aut}(G)$); see Theorem 4.6. Compression techniques can also be used in order to solve the word problem for the outer automorphism group $\mathsf{Out}(G)$ (the quotient of $\mathsf{Aut}(G)$ by the inner automorphism group, which consists of all conjugations $x \mapsto g^1 x g$ for $g \in G$). Schleimer proved in [151] that for a finitely generated free group F, the word problem for $\mathsf{Out}(F)$ can be solved in polynomial time. For this, he proved that a variant of the compressed conjugacy problem for F can be solved in polynomial time, for which he used a polynomial time pattern matching algorithm for compressed words; cf. Sect. 3.4. In [74], Schleimer's result was generalized to graph groups and graph products.

M. Lohrey, *The Compressed Word Problem for Groups*, SpringerBriefs in Mathematics,
DOI 10.1007/978-1-4939-0748-9_7, © Markus Lohrey 2014

In the proofs of Theorems 4.6–4.9, we basically used straight-line programs as a compact representation of long words that occur as intermediate data structures in algorithms. This leads to the idea of using other specialized succinct data structures to store intermediate results. In the context of word problems, we mentioned power circuits in Sect. 2.5. These are succinct representations of huge integers (towers of exponents). Power circuits were designed to solve the word problem in groups like the Baumslag-Gersten group $BG(1,2) = \langle a,b,t \mid bab^{-1} = t, tat^{-1} = a^2 \rangle$ or Higman's group $H_4 = \langle a_0, a_1, a_2, a_3 \mid a_1 a_0 a_1^{-1} = a_0^2, a_2 a_1 a_2^{-1} = a_1^2, a_3 a_2 a_3^{-1} = a_2^2, a_0 a_3 a_0^{-1} = a_3^2 \rangle$. In these groups, standard algorithms for HNN-extensions (Britton reduction) yield big powers a^n, and power circuits can be used to store the exponent n succinctly; see [54, 132]. An interesting project for future work might be to combine SLPs with power circuits, so that words of huge length (and not just singly exponential length as for SLPs) can be produced.

Apart from the solution of word problems, the formalism of straight-line programs turned out to be useful also for other algebraic decision problems, in particular for the solution of word equations. A *word equation* over a finitely generated monoid M (generated by the finite set Γ) is a pair of the form (U, V), where $U, V \in (\Gamma \cup X)^*$. Here, X is a finite set of variables. This word equation is *solvable* if there exists a mapping $\sigma : X \to \Gamma^*$ such that $\sigma(U) = \sigma(V)$ in M, where we extend σ to a morphism on $(\Gamma \cup X)^*$ by setting $\sigma(a) = a$ for $a \in \Gamma$. The mapping σ is called a *solution* of (U, V). In his seminal paper [123], Makanin proved that one can decide whether a given word equation over a free monoid is solvable, and in [140], he extended this result to free groups. Since the work of Makanin, the upper bound on the complexity of solvability of word equations (in free monoids and free groups) was improved several times. An important step was done by Plandowski and Rytter in [141]. To state their result, we need a few definitions (we always refer to equations in free monoids below). Let σ be a solution for a word equation (U, V). We say that σ is *minimal* if, for every solution σ' of (U, V), we have $|\sigma(U)| \leq |\sigma'(U)|$. We say that $|\sigma(U)|$ is the length of a minimal solution of (U, V).

Theorem 7.1 ([141]). *Let (U, V) be a word equation and let $n = |UV|$. Assume that (U, V) has a solution and let N be the length of a minimal solution of (U, V). Then, for every minimal solution σ of (U, V), the word $\sigma(U)$ can be generated by an SLP of size $O(n^2 \log^2(N)(\log n + \log \log N))$.*

Thus, minimal solutions for word equations are highly compressible.

In combination with other ingredients, Plandowski used Theorem 7.1 to show that solvability of word equations over a free monoid belongs to **PSPACE** [140]. This is the currently best known upper bound. Solvability of word equations in a free monoid is easily seen to be **NP**-hard, and it has been repeatedly conjectured that the precise complexity is **NP** too. In [91], Jeż applied his recompression technique from [88, 89] (see Sect. 3.4) to word equations and obtained an alternative **PSPACE** algorithm for solving word equations over a free monoid. Gutiérrez [70] proved that solvability of word equations over free groups belongs to **PSPACE** as well. Further results regarding the compressibility of solutions of word equations can be found in [55].

References

1. S.I. Adjan, The unsolvability of certain algorithmic problems in the theory of groups. Trudy Moskov. Mat. Obsc. **6**, 231–298 (1957) (in Russian)
2. I. Agol, The virtual Haken conjecture. Documenta Math. **18**, 1045–1087 (2013)
3. M. Agrawal, S. Biswas, Primality and identity testing via chinese remaindering. J. Assoc. Comput. Mach. **50**(4), 429–443 (2003)
4. M. Agrawal, N. Kayal, N. Saxena, PRIMES is in P. Ann. Math. Sec. Ser. **160**(2), 781–793 (2004)
5. A. Aho, J.E. Hopcroft, J.D. Ullman, *The Design and Analysis of Computer Algorithms* (Addison–Wesley, Reading, 1974)
6. E. Allender, P. Bürgisser, J. Kjeldgaard-Pedersen, P.B. Miltersen, On the complexity of numerical analysis. SIAM J. Comput. **38**(5), 1987–2006 (2009)
7. S. Arora, B. Barak, *Computational Complexity: A Modern Approach* (Cambridge University Press, Cambridge, 2009)
8. E. Artin, Theorie der Zöpfe. Abhandlungen aus dem Mathematischen Seminar der Universität Hamburg **4**(1), 47–72 (1925)
9. L. Auslander, G. Baumslag, Automorphism groups of finitely generated nilpotent groups. Bull. Amer. Math. Soc. **73**, 716–717 (1967)
10. J. Avenhaus, K. Madlener, Algorithmische Probleme bei Einrelatorgruppen und ihre Komplexität. Archiv für Mathematische Logik und Grundlagenforschung **19**(1–2), 3–12 (1978/79)
11. D.A.M. Barrington, Bounded-width polynomial-size branching programs recognize exactly those languages in NC^1. J. Comput. Syst. Sci. **38**, 150–164 (1989)
12. G. Baumslag, Wreath products and finitely presented groups. Math. Z. **79**, 22–28 (1961)
13. G. Baumslag, A non-cyclic one-relator group all of whose finite quotients are cyclic. J. Austr. Math. Soc. Ser. A. Pure Math. Stat. **10**, 497–498 (1969)
14. G. Baumslag, F.B. Cannonito, C.F. Miller III, Infinitely generated subgroups of finitely presented groups I. Math. Z. **153**(2), 117–134 (1977)
15. M. Beaudry, P. McKenzie, P. Péladeau, D. Thérien, Finite monoids: from word to circuit evaluation. SIAM J. Comput. **26**(1), 138–152 (1997)
16. M. Ben-Or, R. Cleve, Computing algebraic formulas using a constant number of registers. SIAM J. Comput. **21**(1), 54–58 (1992)
17. E. Berlekamp, *Algebraic Coding Theory* (McGraw-Hill, New York, 1968)
18. J.-C. Birget, The groups of Richard Thompson and complexity. Int. J. Algebra Comput. **14**(5–6), 569–626 (2004)
19. J.-C. Birget, A.Y. Ol'shanskii, E. Rips, M.V. Sapir, Isoperimetric functions of groups and computational complexity of the word problem. Ann. Math. Sec. Ser. **156**(2), 467–518 (2002)
20. A. Björner, F. Brenti, *Combinatorics of Coxeter Groups*. Graduate Texts in Mathematics, vol. 231 (Springer, New York, 2005)

21. R.V. Book, F. Otto, *String–Rewriting Systems* (Springer, New York, 1993)
22. W.W. Boone, The word problem. Ann. Math. Sec. Ser. **70**, 207–265 (1959)
23. W.W. Boone, G. Higman, An algebraic characterization of groups with soluble word problem. J. Austr. Math. Soc. **18**, 41–53 (1974)
24. B. Brink, R.B. Howlett, A finiteness property and an automatic structure for Coxeter groups. Math. Ann. **296**, 179–190 (1993)
25. J.-Y. Cai, Parallel computation over hyperbolic groups. In *Proceedings of the 24th Annual Symposium on Theory of Computing, STOC 1992* (ACM Press, New York, 1992), pp. 106–115
26. J.W. Cannon, W.J. Floyd, W.R. Parry, Introductory notes on Richard Thompson's groups. L'Enseignement Mathématique **42**(3), 215–256 (1996)
27. F.B. Cannonito, Hierarchies of computable groups and the word problem. J. Symbolic Logic **31**, 376–392 (1966)
28. F.B. Cannonito, R.W. Gatterdam, The word problem in polycyclic groups is elementary. Compos. Math. **27**, 39–45 (1973)
29. F.B. Cannonito, R.W. Gatterdam, The word problem and power problem in 1-relator groups are primitive recursive. Pacific J. Math. **61**(2), 351–359 (1975)
30. W.A. Casselman, Automata to perform basic calculations in coxeter groups. *C.M.S. Conference Proceedings*, vol. 16, 1994
31. M. Charikar, E. Lehman, A. Lehman, D. Liu, R. Panigrahy, M. Prabhakaran, A. Sahai, A. Shelat, The smallest grammar problem. IEEE Trans. Inform. Theor. **51**(7), 2554–2576 (2005)
32. R. Charney, An introduction to right-angled artin groups. Geometriae Dedicata **125**, 141–158 (2007)
33. R. Charney, K. Vogtmann, Finiteness properties of automorphism groups of right-angled artin groups. Bull. Lond. Math. Soc. **41**(1), 94–102 (2009)
34. R. Charney, J. Crisp, K. Vogtmann, Automorphisms of 2-dimensional right-angled artin groups. Geom. Topology **11**, 2227–2264 (2007)
35. A. Chiu, G. Davida, B. Litow, Division in logspace-uniform NC^1. Theor. Inform. Appl. Inform. Théor. Appl. **35**(3), 259–275 (2001)
36. W.H. Cockcroft, The word problem in a group extension. Q. J. Math. Oxford Sec. Ser. **2**, 123–134 (1951)
37. D.E. Cohen, String rewriting—a survey for group theorists. In *Geometric Group Theory, Vol. 1 (Sussex, 1991)*. London Mathematical Society Lecture Note Series, vol. 181 (Cambridge University Press, Cambridge, 1993), pp. 37–47
38. D.E. Cohen, K. Madlener, F. Otto, Separating the intrinsic complexity and the derivational complexity of the word problem for finitely presented groups. Math. Logic Q. **39**(2), 143–157 (1993)
39. S.A. Cook, The complexity of theorem–proving procedures. In *Proceedings of the 3rd Annual ACM Symposium on Theory of Computing, STOC 1971*, pp. 151–158, 1971
40. B. Cooper, *Computability Theory* (Chapman & Hall/CRC Mathematics, Boca Raton, 2003)
41. R. Cori, Y. Métivier, W. Zielonka, Asynchronous mappings and asynchronous cellular automata. Inform. Comput. **106**(2), 159–202 (1993)
42. T.H. Cormen, C.E. Leiserson, R.L. Rivest, C. Stein, *Introduction to Algorithms*, 3rd edn. (MIT Press, Cambridge, 2009)
43. L.J. Corredor, M.A. Gutierrez, A generating set for the automorphism group of a graph product of abelian groups. Int. J. Algebra Comput. **22**(1), 1250003, 21 p (2012)
44. M.B. Day, Peak reduction and finite presentations for automorphism groups of right-angled artin groups. Geom. Topology **13**(2), 817–855 (2009)
45. M. Dehn, Über die toplogie des dreidimensionalen raumes. Math. Ann. **69**, 137–168 (1910) (in German)
46. M. Dehn, Über unendliche diskontinuierliche gruppen. Math. Ann. **7**, 116–144 (1911) (in German)

47. M. Dehn, Transformation der Kurven auf zweiseitigen Flächen. Math. Ann. **72**, 413–421 (1912) (in German)
48. R.A. DeMillo, R.J. Lipton, A probabilistic remark on algebraic program testing. Inform. Process. Lett. **7**(4), 193–195 (1978)
49. V. Diekert, *Combinatorics on Traces*. Lecture Notes in Computer Science, vol. 454 (Springer, New York, 1990)
50. V. Diekert, J. Kausch, Logspace computations in graph products. Technical Report. arXiv.org (2013), http://arxiv.org/abs/1309.1290
51. V. Diekert, M. Lohrey, Word equations over graph products. Int. J. Algebra Comput. **18**(3), 493–533 (2008)
52. V. Diekert, G. Rozenberg (eds.), *The Book of Traces* (World Scientific, Singapore, 1995)
53. V. Diekert, A.J. Duncan, A.G. Myasnikov, Geodesic rewriting systems and pregroups. In *Combinatorial and Geometric Group Theory*. Trends in Mathematics (Birkhäuser, Boston, 2010), pp. 55–91
54. V. Diekert, J. Laun, A. Ushakov, Efficient algorithms for highly compressed data: the word problem in Higman's group is in P. Int. J. Algebra Comput. **22**(8), 19 p (2012)
55. V. Diekert, O. Kharlampovich, A.M. Moghaddam, SLP compression for solutions of equations with constraints in free and hyperbolic groups. Technical Report. arXiv.org (2013), http://arxiv.org/abs/1308.5586
56. C. Droms, A complex for right-angled Coxeter groups. Proc. Amer. Math. Soc. **131**(8), 2305–2311 (2003)
57. D.B.A. Epstein, J.W. Cannon, D.F. Holt, S.V.F. Levy, M.S. Paterson, W.P. Thurston, *Word Processing in Groups* (Jones and Bartlett, Boston, 1992)
58. M.R. Garey, D.S. Johnson, *Computers and Intractability: A Guide to the Theory of NP–completeness* (Freeman, New York, 1979)
59. L. Gasieniec, M. Karpinski, W. Plandowski, W. Rytter, Efficient algorithms for Lempel-Ziv encoding (extended abstract). In *Proceedings of the 5th Scandinavian Workshop on Algorithm Theory, SWAT 1996*. Lecture Notes in Computer Science, vol. 1097 (Springer, Berlin, 1996), pp. 392–403
60. L. Gasieniec, M. Karpinski, W. Plandowski, W. Rytter, Randomized efficient algorithms for compressed strings: the finger-print approach (extended abstract). In *Proceedings of the 7th Annual Symposium on Combinatorial Pattern Matching, CPM 96*. Lecture Notes in Computer Science, vol. 1075 (Springer, Berlin, 1996), pp. 39–49
61. S. M. Gersten, Dehn functions and ℓ_1-norms of finite presentations. In *Algorithms and Classification in Combinatorial Group Theory*, ed. by G. Baumslag, C.F. Miller III (Springer, New York, 1991), pp. 195–220
62. S.M. Gersten, D.F. Holt, T.R. Riley, Isoperimetric inequalities for nilpotent groups. Geom. Funct. Anal. **13**(4), 795–814 (2003)
63. L.M. Goldschlager, The monotone and planar circuit value problems are log space complete for P. SIGACT News **9**(2), 25–99 (1977)
64. O. Goodman, M. Shapiro, On a generalization of Dehn's algorithm. Int. J. Algebra Comput. **18**(7), 1137–1177 (2008)
65. K. Goto, H. Bannai, S. Inenaga, M. Takeda, Fast q-gram mining on SLP compressed strings. In *Proceedings of the 18th International Symposium on String Processing and Information Retrieval, SPIRE 2011*. Lecture Notes in Computer Science, vol. 7024 (Springer, Berlin, 2011), pp. 278–289
66. E.R. Green, Graph products of groups. Ph.D. thesis, The University of Leeds, 1990
67. R. Greenlaw, H.J. Hoover, W.L. Ruzzo, *Limits to Parallel Computation: P-Completeness Theory* (Oxford University Press, Oxford, 1995)
68. M. Gromov, Hyperbolic groups. In *Essays in Group Theory*, ed. by S.M. Gersten. MSRI Publ., vol. 8 (Springer, New York, 1987), pp. 75–263
69. V.S. Guba, M.V. Sapir, On subgroups of R. Thompson's group F and other diagram groups. Matematicheskii Sbornik **190**(8), 3–60 (1999)

70. C. Gutiérrez, Satisfiability of equations in free groups is in PSPACE. In *Proceedings of the 32nd Annual ACM Symposium on Theory of Computing, STOC 2000* (ACM Press, New York, 2000), pp. 21–27

71. C. Hagenah, Gleichungen mit regulären Randbedingungen über freien Gruppen. Ph.D. thesis, University of Stuttgart, Institut für Informatik, 2000

72. F. Haglund, D.T. Wise, Coxeter groups are virtually special. Adv. Math. **224**(5), 1890–1903 (2010)

73. N. Haubold, M. Lohrey, Compressed word problems in HNN-extensions and amalgamated products. Theor. Comput. Syst. **49**(2), 283–305 (2011)

74. N. Haubold, M. Lohrey, C. Mathissen, Compressed decision problems for graph products of groups and applications to (outer) automorphism groups. Int. J. Algebra Comput. **22**(8), 53 p (2013)

75. S. Hermiller, J. Meier, Algorithms and geometry for graph products of groups. J. Algebra **171**, 230–257 (1995)

76. W. Hesse, E. Allender, D.A.M. Barrington, Uniform constant-depth threshold circuits for division and iterated multiplication. J. Comput. Syst. Sci. **65**, 695–716 (2002)

77. G. Higman, Subgroups of finitely presented groups. Proc. Roy. Soc. Lond. Ser. A **262**, 455–475 (1961)

78. G. Higman, B.H. Neumann, H. Neumann, Embedding theorems for groups. J. Lond. Math. Soc. Sec. Ser. **24**, 247–254 (1949)

79. Y. Hirshfeld, M. Jerrum, F. Moller, A polynomial-time algorithm for deciding equivalence of normed context-free processes. In *Proceedings of the 35th Annual Symposium on Foundations of Computer Science, FOCS 1994* (IEEE Computer Society, Los Alamitos, 1994), pp. 623–631

80. Y. Hirshfeld, M. Jerrum, F. Moller, A polynomial algorithm for deciding bisimilarity of normed context-free processes. Theor. Comput. Sci. **158**(1&2), 143–159 (1996)

81. D. Holt, Word-hyperbolic groups have real-time word problem. Int. J. Algebra Comput. **10**, 221–228 (2000)

82. J.E. Hopcroft, J.D. Ullman, *Introduction to Automata Theory, Languages and Computation* (Addison–Wesley, Reading, 1979)

83. T. Hsu, D.T. Wise, On linear and residual properties of graph products. Mich. Math. J. **46**(2), 251–259 (1999)

84. O.H. Ibarra, S. Moran, Probabilistic algorithms for deciding equivalence of straight-line programs. J. Assoc. Comput. Mach. **30**(1), 217–228 (1983)

85. R. Impagliazzo, A. Wigderson, P = BPP if E requires exponential circuits: derandomizing the XOR lemma. In *Proceedings of the 29th Annual ACM Symposium on the Theory of Computing, STOC 1997* (ACM Press, New York, 1997), pp. 220–229

86. N. Jacobson, *Lectures in Abstract Algebra, Vol III: Theory of Fields and Galois Theory* (D. Van Nostrand Co. Inc., Princeton, 1964)

87. N. Jacobson, *Basic Algebra I* (W. H. Freeman and Company, New York, 1974)

88. A. Jėz, Compressed membership for NFA (DFA) with compressed labels is in NP (P). In *Proceedings of the 29th International Symposium on Theoretical Aspects of Computer Science, STACS 2012*. Leibniz International Proceedings in Informatics, vol. 14 (Schloss Dagstuhl - Leibniz-Zentrum für Informatik, Germany, 2012), pp. 136–147

89. A. Jėz, Faster fully compressed pattern matching by recompression. In *Proceedings of the 39th International Colloquium on Automata, Languages, and Programming, ICALP 2012*. Lecture Notes in Computer Science, vol. 7391 (Springer, New York, 2012), pp. 533–544

90. A. Jėz, Approximation of grammar-based compression via recompression. In *Proceedings of the 24th Annual Symposium on Combinatorial Pattern Matching, CPM 2013*. Lecture Notes in Computer Science, vol. 7922 (Springer, New York, 2013), pp. 165–176

91. A. Jèz, Recompression: a simple and powerful technique for word equations. In *Proceedings of the 30th International Symposium on Theoretical Aspects of Computer Science, STACS 2013*. Leibniz International Proceedings in Informatics, vol. 20 (Schloss Dagstuhl - Leibniz-Zentrum für Informatik, Germany, 2013), pp. 233–244

92. V. Kabanets, R. Impagliazzo, Derandomizing polynomial identity tests means proving circuit lower bounds. Comput. Complex. **13**(1–2), 1–46 (2004)

93. E. Kaltofen, Greatest common divisors of polynomials given by straight-line programs. J. Assoc. Comput. Mach. **35**(1), 231–264 (1988)

94. I. Kapovich, A. Miasnikov, P. Schupp, V. Shpilrain, Generic-case complexity, decision problems in group theory, and random walks. J. Algebra **264**(2), 665–694 (2003)

95. M.I. Kargapolov, J.I. Merzljakov, *Fundamentals of the Theory of Groups*. Graduate Texts in Mathematics, vol. 62 (Springer, New York, 1979)

96. M. Karpinski, W. Rytter, A. Shinohara, Pattern-matching for strings with short descriptions. In *Proceedings of the 6th Annual Symposium on Combinatorial Pattern Matching, CPM 95*. Lecture Notes in Computer Science, vol. 937 (Springer, New York, 1995), pp. 205–214

97. O. Kharlampovich, A. Myasnikov, M. Sapir, Algorithmically complex residually finite groups. Technical Report. arXiv.org (2012), http://arxiv.org/abs/1204.6506

98. T. Kida, T. Matsumoto, Y. Shibata, M. Takeda, A. Shinohara, S. Arikawa, Collage system: a unifying framework for compressed pattern matching. Theor. Comput. Sci. **298**(1), 253–272 (2003)

99. D.E. Knuth, J.H. Morris Jr., V.R. Pratt, Fast pattern matching in strings. SIAM J. Comput. **6**(2), 323–350 (1977)

100. D. Kuske, M. Lohrey, Logical aspects of Cayley-graphs: the monoid case. Int. J. Algebra Comput. **16**(2), 307–340 (2006)

101. R.E. Ladner, The circuit value problem is log space complete for P. SIGACT News **7**(1), 18–20 (1975)

102. M.R. Laurence, A generating set for the automorphism group of a graph group. J. Lond. Math. Soc. Sec. Ser. **52**(2), 318–334 (1995)

103. M.V. Lawson, *Inverse Semigroups: The Theory of Partial Symmetries* (World Scientific, Singapore, 1999)

104. L. Levin, Universal search problems. Probl. Inform. Transm. **9**(3), 265–266 (1973) (in Russian)

105. J. Lewin, Residual properties of loops and rings. Ph.D. thesis, New York University, 1964

106. Y. Lifshits, Processing compressed texts: a tractability border. In *Proceedings of the 18th Annual Symposium on Combinatorial Pattern Matching, CPM 2007*. Lecture Notes in Computer Science, vol. 4580 (Springer, New York, 2007), pp. 228–240

107. Y. Lifshits, M. Lohrey, Querying and embedding compressed texts. In *Proceedings of the 31th International Symposium on Mathematical Foundations of Computer Science, MFCS 2006*. Lecture Notes in Computer Science, vol. 4162 (Springer, Berlin, 2006), pp. 681–692

108. R.J. Lipton, Y. Zalcstein, Word problems solvable in logspace. J. Assoc. Comput. Mach. **24**(3), 522–526 (1977)

109. H.-N. Liu, C. Wrathall, K. Zeger, Efficient solution to some problems in free partially commutative monoids. Inform. Comput. **89**(2), 180–198 (1990)

110. M. Lohrey, Word problems on compressed word. In *Proceedings of the 31st International Colloquium on Automata, Languages and Programming, ICALP 2004*. Lecture Notes in Computer Science, vol. 3142 (Springer, Berlin, 2004), pp. 906–918

111. M. Lohrey, Decidability and complexity in automatic monoids. Int. J. Found. Comput. Sci. **16**(4), 707–722 (2005)

112. M. Lohrey, Word problems and membership problems on compressed words. SIAM J. Comput. **35**(5), 1210–1240 (2006)

113. M. Lohrey, Compressed word problems for inverse monoids. In *Proceedings of the 36th International Symposium on Mathematical Foundations of Computer Science, MFCS 2011*. Lecture Notes in Computer Science, vol. 6907 (Springer, Berlin, 2011), pp. 448–459

114. M. Lohrey, Algorithmics on SLP-compressed strings: a survey. Groups Complex. Cryptol.
     4(2), 241–299 (2012)
115. M. Lohrey, N. Ondrusch, Inverse monoids: decidability and complexity of algebraic questions. Inform. Comput. **205**(8), 1212–1234 (2007)
116. M. Lohrey, S. Schleimer, Efficient computation in groups via compression. In *Proceedings of Computer Science in Russia, CSR 2007*. Lecture Notes in Computer Science, vol. 4649 (Springer, Berlin, 2007), pp. 249–258
117. M. Lohrey, G. Sénizergues, Theories of HNN-extensions and amalgamated products. In *Proceedings of the 33st International Colloquium on Automata, Languages and Programming, ICALP 2006*. Lecture Notes in Computer Science, vol. 4052 (Springer, Berlin, 2006), pp. 681–692
118. M. Lohrey, B. Steinberg, G. Zetzsche, Rational subsets and submonoids of wreath products. Technical Report. arXiv.org (2013), http://arxiv.org/abs/1302.2455
119. R.C. Lyndon, P.E. Schupp, *Combinatorial Group Theory* (Springer, New York, 1977)
120. J. Macdonald, Compressed words and automorphisms in fully residually free groups. Int. J. Algebra Comput. **20**(3), 343–355 (2010)
121. W. Magnus, Das Identitätsproblem für gruppen mit einer definierenden relation. Math. Ann. **106**(1), 295–307 (1932)
122. W. Magnus, On a theorem of Marshall hall. Ann. Math. Sec. Ser. **40**, 764–768 (1939)
123. G.S. Makanin, The problem of solvability of equations in a free semigroup. Math. Sbornik **103**, 147–236 (1977). In Russian; English translation in Math. USSR Sbornik 32, 1977.
124. N. Markey, P. Schnoebelen, A PTIME-complete matching problem for SLP-compressed words. Inform. Process. Lett. **90**(1), 3–6 (2004)
125. J. Matthews, The conjugacy problem in wreath products and free metabelian groups. Trans. Amer. Math. Soc. **121**, 329–339 (1966)
126. J. McKinsey, The decision problem for some classes of sentences without quantifiers. J. Symbolic Logic **8**, 61–76 (1973)
127. K. Mehlhorn, R. Sundar, C. Uhrig, Maintaining dynamic sequences under equality-tests in polylogarithmic time. In *Proceedings of the 5th Annual ACM-SIAM Symposium on Discrete Algorithms, SODA 1994* (ACM/SIAM, New York, 1994), pp. 213–222
128. K. Mehlhorn, R. Sundar, C. Uhrig, Maintaining dynamic sequences under equality tests in polylogarithmic time. Algorithmica **17**(2), 183–198 (1997)
129. J. Meier, When is the graph product of hyperbolic groups hyperbolic? Geometriae Dedicata **61**, 29–41 (1996)
130. M. Miyazaki, A. Shinohara, M. Takeda, An improved pattern matching algorithm for strings in terms of straight-line programs. In *Proceedings of the 8th Annual Symposium on Combinatorial Pattern Matching, CPM 97*. Lecture Notes in Computer Science, vol. 1264 (Springer, Berlin, 1997), pp. 1–11
131. A. Myasnikov, V. Shpilrain, A. Ushakov, in *Non-commutative Cryptography and Complexity of Group-theoretic Problems*. Mathematical Surveys and Monographs, vol. 177 (American Mathematical Society, Providence, 2011)
132. A.G. Myasnikov, A. Ushakov, D.W. Won, The word problem in the Baumslag group with a non-elementary Dehn function is polynomial time decidable. J. Algebra **345**(1), 324–342 (2011)
133. A.G. Myasnikov, A. Ushakov, D.W. Won, Power circuits, exponential algebra, and time complexity. Int. J. Algebra Comput. **22**(6), (2012)
134. M.H.A. Newman, On theories with a combinatorial definition of "equivalence". Ann. Math. **43**, 223–243 (1943)
135. J. Nielsen, Die Isomorphismengruppe der freien gruppen. Math. Ann. **91**(3–4), 169–209 (1924)
136. P.S. Novikov, On the algorithmic unsolvability of the word problem in group theory. Amer. Math. Soc. Transl. II. Ser. **9**, 1–122 (1958)
137. A.Y. Ol'shanskiĭ, Almost every group is hyperbolic. Int. J. Algebra Comput. **2**(1), 1–17 (1992)

138. C.H. Papadimitriou, *Computational Complexity* (Addison Wesley, Reading, 1994)
139. W. Plandowski, Testing equivalence of morphisms on context-free languages. In *Proceedings of the 2nd Annual European Symposium on Algorithms, ESA 1994*. Lecture Notes in Computer Science, vol. 855 (Springer, New York, 1994), pp. 460–470
140. W. Plandowski, Satisfiability of word equations with constants is in PSPACE. J. Assoc. Comput. Mach. **51**(3), 483–496 (2004)
141. W. Plandowski, W. Rytter, Application of Lempel-Ziv encodings to the solution of word equations. In *Proceedings of the 25th International Colloquium on Automata, Languages and Programming, ICALP 1998*. Lecture Notes in Computer Science, vol. 1443 (Springer, Berlin, 1998), pp. 731–742
142. W. Plandowski, W. Rytter, Complexity of language recognition problems for compressed words. In *Jewels are Forever, Contributions on Theoretical Computer Science in Honor of Arto Salomaa*, ed. by J. Karhumäki, H.A. Maurer, G. Paun, G. Rozenberg (Springer, Berlin, 1999), pp. 262–272
143. A.N. Platonov, An isoparametric function of the Baumslag-Gersten group. Vestnik Moskov. Univ. Ser. I Mat. Mekh. **3**, 12–17, 70 p (2004)
144. M.O. Rabin, Recursive unsolvability of group theoretic problems. Ann. Math. Sec. Ser. **67**, 172–194 (1958)
145. M. Rabin, Computable algebra, general theory and theory of computable fields. Trans. Amer. Math. Soc. **95**, 341–360 (1960)
146. O. Reingold, Undirected st-connectivity in log-space. In *Proceedings of the 37th Annual ACM Symposium on Theory of Computing, STOC 2005* (ACM Press, New York, 2005), pp. 376–385
147. T. Riley, What is a Dehn function? In *Office Hours with a Geometric Group Theorist*, ed. by M. Clay, D. Margalit (Princeton University Press, Princeton, to appear)
148. D. Robinson, Parallel algorithms for group word problems. Ph.D. thesis, University of California, San Diego, 1993
149. J.J. Rotman, *An Introduction to the Theory of Groups*, 4th edn. (Springer, New York, 1995)
150. W. Rytter, Grammar compression, LZ-encodings, and string algorithms with implicit input. In *Proceedings of the 31st International Colloquium on Automata, Languages and Programming, ICALP 2004*. Lecture Notes in Computer Science, vol. 3142 (Springer, New York, 2004), pp. 15–27
151. S. Schleimer, Polynomial-time word problems. Comment. Math. Helv. **83**(4), 741–765 (2008)
152. J.T. Schwartz, Fast probabilistic algorithms for verification of polynomial identities. J. Assoc. Comput. Mach. **27**(4), 701–717 (1980)
153. J.-P. Serre, *Trees* (Springer, Berlin, 2003)
154. H. Servatius, Automorphisms of graph groups. J. Algebra **126**(1), 34–60 (1989)
155. H.-U. Simon, Word problems for groups and contextfree recognition. In *Proceedings of Fundamentals of Computation Theory, FCT 1979* (Akademie, Berlin, 1979), pp. 417–422
156. J.R. Stallings, *Group Theory and Three-Dimensional Manifolds*. Yale Mathematical Monographs, vol. 4 (Yale University Press, London, 1971)
157. J. Stillwell, The word problem and the isomorphism problem for groups. Bull. Amer. Math. Soc. **6**(1), 33–56 (1982)
158. J. Stillwell, *Classical Topology and Combinatorial Group Theory*, 2nd edn. (Springer, New York, 1995)
159. A. Thue, Probleme über die Veränderungen von Zeichenreihen nach gegebenen Regeln. Skr. Vid. Kristiania, I Math. Natuv. Klasse, No. 10, 34 S, (1914) (in German)
160. H. Tietze, Über die topologischen Invarianten mehrdimensionaler Mannigfaltigkeiten. Monatsh. Math. Phys. **19**(1), 1–118 (1908)
161. A.M. Turing, On computable numbers, with an application to the Entscheidungsproblem. Proc. Lond. Math. Soc. Ser. 2 **42**, 230–265 (1937)
162. H. Vollmer. *Introduction to Circuit Complexity* (Springer, New York, 1999)
163. S. Waack, The parallel complexity of some constructions in combinatorial group theory. J. Inform. Process. Cybern. EIK **26**, 265–281 (1990)

164. B.A. Wehrfritz, On finitely generated soluble linear groups. Math. Z. **170**, 155–167 (1980)
165. D.T. Wise, Research announcement: the structure of groups with a quasiconvex hierarchy. Electron. Res. Announc. Math. Sci. **16**, 44–55 (2009)
166. C. Wrathall, The word problem for free partially commutative groups. J. Symbolic Comput. **6**(1), 99–104 (1988)
167. R. Zippel, Probabilistic algorithms for sparse polynomials. In *EUROSAM*, ed. by E.W. Ng. Lecture Notes in Computer Science, vol. 72 (Springer, New York, 1979), pp. 216–226

Acronyms and Notations

$\leftrightarrow^*$	Equivalence relation generated by $\rightarrow$, see p. 2
$\rightarrow_R$	One-step rewrite relation, see pp. 3 and 92
$\vdash_M$	Transition relation between configurations of the Turing machine M, see p. 6
$\leq_m^P$	Polynomial time many-one reducible, see p. 11
$\leq_m^{\log}$	Logspace many-one reducible, see p. 11
$\leq_{bc}^P$	Polynomial time bounded conjunctively reducible, see p. 11
$\leq_{bc}^{\log}$	Logspace bounded conjunctively reducible, see p. 11
$\leq_T^P$	Polynomial time Turing-reducible, see p. 12
$\preceq_p$	Prefix order on traces, see p. 91
$\preceq_s$	Suffix order on traces, see p. 91
$[a, b]$	Commutator of a and b, see p. 28
$\langle B \rangle$	Subgroup generated by B, see p. 28
$\langle B \rangle^G$	Normal closure of $B \subseteq G$, see p. 28
$H \wr G$	Wreath product of H and G, see p. 83
$K \rtimes_\varphi Q$	Semidirect product of K and Q, see p. 70
$|s|$	Length of the word s, see p. 1
$|s|_a$	Number of occurrences of symbol a in the word s, see p. 1
$s[i]$	Symbol at position i in s, see p. 1
$s[i : j]$	Factor of s from position i to position j, see p. 1
$s[i :]$	Suffix of s starting at position i, see p. 1
$s[: j]$	Prefix of s ending at position i, see p. 1
$[s]_R$	Equivalence class of s with respect to $\leftrightarrow_R^*$, see p. 3
$s[\Sigma_l, \Sigma_r]$	See p. 55

$[\![\mathscr{T}]\!]$	Mapping computed by the deterministic rational transducer $\mathscr{T}$, see p. 48
$u \sqcap_p v$	Prefix infimum of the traces u and v, see p. 92
$u \sqcap_s v$	Suffix infimum of the traces u and v, see p. 92
$u \setminus_p v$	Unique trace w such that $u = (u \sqcap_p v)w$, see p. 92
$u \setminus_s v$	Unique trace w such that $u = w(u \sqcap_s v)$, see p. 92
$[w]_I$	Trace represented by the word w, see p. 90
Γ^*	Set of finite words over Γ and free monoid generated by Γ, see p. 1
Γ^+	Set of finite nonempty words over Γ, see p. 1
Γ^*/R	Quotient of Γ^* by the semi-Thue system R, see p. 3
$\langle \Gamma \mid R \rangle$	Group presented by (Γ, R), see p. 28
$[\Delta]_I$	Trace represented by the independence clique Δ, see p. 90
ε	Empty word, see p. 1
π_Δ	Projection homomorphism, see p. 43 and 91
accept	Set of accepting configurations, see p. 5
alph(s)	Set of letters that occur in the word s, see p. 1
Aut(G)	Automorphism group of G, see p. 70
block(s)	See p. 55
BPP	Bounded error probabilistic polynomial time, see p. 19
C-expression	See p. 44
$C(M)$	Coxeter group defined by the matrix M, see p. 29
char(R)	Characteristic of the ring R, see p. 20
coC	Set of all complements of languages from the class **C**, see p. 10
coRP	Co-randomized polynomial time, see p. 18
CSLP	Straight-line program with cut, see p. 44
CWP(G)	Compressed word problem for the group G, see p. 68
CWP(G, Γ)	Compressed word problem for the group G with respect to the generating set Γ, see p. 67
$D(a)$	Set of letter not commuting with a, see p. 90
$D_{\Gamma,R}$	Dehn function for the group presentation (Γ, R), see p. 38
D_w	Dependence graph for the word w, see p. 90
DFA	Deterministic finite automaton, see p. 2
DSPACE(f)	Class of languages that can be decided by an f-space bounded deterministic Turing machine, see p. 8
DTIME(f)	Class of languages that can be decided by an f-time bounded deterministic Turing machine, see p. 8
$E[C_1, \ldots, C_n]$	See p. 93
$F(\Gamma)$	Free group generated by Γ, see p. 27
$G(A, I)$	Graph group defined by the graph (A, I), see p. 29

$\mathbb{G}(W, E, (G_i)_{i \in W})$	Graph product, see p. 88
$GL_d(F)$	General linear group of dimension d over the field F, see p. 33
$graph(\mathscr{C})$	Edge relation of the circuit $\mathscr{C}$, see p. 14
$I(a)$	Set of letter commuting with a, see p. 90
$init(x)$	Initial configuration for input x, see p. 5
$IRR(\rightarrow)$	Set of irreducible elements with respect to the relation $\rightarrow$, see p. 3
$IRR(R)$	Set of irreducible words or traces with respect to R, see p. 3 and 92
L	Deterministic logspace, see p. 9
$L(\mathscr{A})$	Language accepted by the NFA $\mathscr{A}$, see p. 2
$L(M)$	Language accepted by the Turing machine M, see p. 6
$lo(\mathbb{A})$	Lower part of the 2-level PCSLP $\mathbb{A}$, see p. 64
$\mathbb{M}(\Sigma, I)$	Trace monoid defined by (Σ, I), see p. 89
$max(u)$	Maximal symbols of the trace u, see p. 92
$min(u)$	Minimal symbols of the trace u, see p. 92
$NF_{\rightarrow}(a)$	Normal form of a with respect to the relation $\rightarrow$, see p. 3
$NF_R(s)$	Normal form of s with respect to the semi-Thue system or trace rewrite system R, see pp. 3 and 93
NFA	Nondeterministic finite automaton, see p. 2
NL	Nondeterministic logspace, see p. 9
NP	Nondeterministic polynomial time, see p. 9
$NSPACE(f)$	Class of languages that can be decided by an f-space bounded nondeterministic Turing machine, see p. 8
$NTIME(f)$	Class of languages that can be decided by an f-time bounded nondeterministic Turing machine, see p. 8
P	Deterministic polynomial time, see p. 9
$\mathscr{P}_j$	Simple paths in a dependence alphabet that start at node j, see p. 105
PC-expression	See p. 44
PCSLP	Straight-line program with projection and cut, see p. 44
$PIT(R)$	Polynomial identity testing for the ring R, see p. 21
$pos(s, J, i)$	See p. 104
PSPACE	Polynomial space, see p. 9
RAM	Random access machine, see p. 25

RCWP$(H, A, B, \varphi_1, \ldots, \varphi_n)$	See p. 120
Red$(H, \varphi_1, \ldots, \varphi_n)$	Set of Britton-reduced words, see p. 116
rhs	Right-hand side mapping of a circuit, see p. 14
RP	Randomized polynomial time, see p. 18
RUCWP(H, A, B)	See p. 120
SAT	Boolean satisfiability problem, see p. 17
SL$_d (\mathbb{Z})$	Special linear group of dimension d over the ring $\mathbb{Z}$, see p. 79
SLP	Straight-line program, see p. 44
SUBSETSUM	Subset sum problem, see p. 18
UCWP(H, A, B)	See p. 119
up$(\mathbb{A})$	Upper part of the 2-level PCSLP $\mathbb{A}$, see p. 64
val$_{\mathscr{C}}^{\mathscr{A}}$	Evaluation mapping of the circuit $\mathscr{C}$ in the algebra $\mathscr{A}$, see p. 15
Var(e)	Variables that occur in the expression e, see p. 14
WP(G)	Word problem for the group G, see p. 32
WP(G, Γ)	Word problem for the group G with respect to the generating set Γ, see p. 32
WSP(G)	Word search problem for the group G, see p. 41
WSP(G, Γ)	Word search problem for the group G with respect to the generating set Γ, see p. 39
ZPP	Zero-error probabilistic polynomial time, see p. 19

Index

Symbols
2-level PCSLP, 63

A
accepting configurations, 5
algebra, 15
alphabet, 1
alphabet of a word, 1
amalgamated free product, 134
area, 36
arithmetic circuit, 20
automatic group, 29
automatic structure for group, 29
automorphism group, 70

B
boolean circuit, 17
bounded error probabilistic polynomial time, 19
Britton-reduced word, 116

C
C-expression, 44
characteristic of the ring R, 20
Chomsky normal form, 45
circuit, 14
circuit in normal form, 15
circuit value problem, 17
co-randomized polynomial time, 18
commutator, 28
complete language, 12
compressed word problem, 67
computable language, 8
computably enumerable, 6

computation of a Turing machine, 6
concatenation of words, 1
configuration, 5
confluent, 3
connecting elements, 117
convex set, 91
convolution of words, 29
Coxeter group, 29
Coxeter matrix, 29
CSLP, 44

D
decidable language, 8
Dehn function, 38
dependence alphabet, 87
dependence graphs, 90
derivation tree, 45
deterministic finite automaton, 2
deterministic rational transducer, 48
deterministic Turing machine, 5
DFA, 2
downward-closed set, 91

E
empty word, 1
expression, 14

F
factor of a word, 1
finite word, 1
finitely generated group, 28
finitely presented group, 28
free group, 27
free monoid, 2

free partially commutative grou, 29
free product of groups, 30

G
general linear group, 33
graph accessibility problem, 16
graph group, 29
graph product, 88

H
hard language, 12
height of a circuit, 14
hierarchical order of a circuit, 14
HNN-extension, 116

I
independence alphabet, 87
independence clique, 90
initial configuration, 5
irreducible element with respect to $\rightarrow$, 3
irreducible trace with respect to trace rewriting
    system, 92
irreducible word in free group, 27
irreducible word with respect to semi-Thue
    system, 3
irredundant 2-level PCSLP, 99

L
language, 2
length of a word, 1
Levi's Lemma, 90
locally confluent, 3
logspace bounded conjunctively reducible, 11
logspace many-one reducibility, 11
logspace transducer, 10
lower level variables, 64
lower central series, 80
lower part of a 2-level PCSLP, 64

M
Magnus embedding theorem, 85
metabelian, 35
monotone circuit value problem, 17

N
Newman's lemma, 3
NFA, 2
nicely projecting 2-level PCSLP, 99

nilpotent group, 80
Noetherian, 3
nondeterministic finite automaton, 2
nondeterministic Turing machine, 4
normal closure, 28
normal form of a trace, 93
normal form of a word, 3

O
one-relator group, 35
one-step rewrite relation, 3, 92
oracle Turing machine, 12

P
PC-expression, 44
PCSLP, 44
polynomial identity testing, 21
polynomial time bounded conjunctively
    reducible, 11
polynomial time many-one reducibility, 11
polynomial time transducer, 10
polynomial time Turing-reducible, 12
prefix infimum, 92
prefix of a trace, 91
presentation of a group, 28
PRIMES, 17
problem, 8
projection homomorphism, 43, 91
pure 2-level PCSLP, 99

Q
quotient monoid, 3

R
RAM, 25
random access machine, 25
randomized polynomial time, 18
recompression technique, 55
recursive language, 8
recursively enumerable, 6
regular language, 2
relator of a group, 28
residually-finite group, 35
right-angled Artin group, 29

S
SAT, 17
saturated 2-level PCSLP, 99
semi-Thue system, 3

semidirect product, 70
simple path, 105
size of a circuit, 15
SLP, 44
space bounded, 8
special linear group, 79
SUBSETSUM, 18
successful computation, 6
suffix infimum, 92
suffix of a trace, 91

T
terminating, 3
Tietze transformations, 30
time bounded, 8
trace monoid, 89
trace rewriting system, 92
transducer, 7
Turing machine with output, 7

U
undecidable language, 8

undirected graph accessibility problem, 17
unitriangular matrix, 81
univariate arithmetic circuit, 20
universality problem for nondeterministic finite
          automata, 18
upper part of a 2-level PCSLP, 64
upper level variables, 64

V
van Kampen diagram, 36
variable-free arithmetic circuit, 20

W
well-formed 2-level PCSLP, 100
word equation, 138
word problem, 32
word search problem, 39
wreath product, 83

Z
zero-error probabilistic polynomial time, 19